T0221200

Information and Instruct

This shop manual contains several sections each covering a specific group of wheel type tractors. The Tab Index on the preceding page can be used to locate the section pertaining to each group of tractors. Each section contains the necessary specifications and the brief but terse procedural data needed by a mechanic when repairing a tractor on which he has had no previous actual experience.

Within each section, the material is arranged in a systematic order beginning with an index which is followed immediately by a Table of Condensed Service Specifications. These specifications include dimensions, fits, clearances and timing instructions. Next in order of arrangement is the procedures paragraphs.

In the procedures paragraphs, the order of presentation starts with the front axle system and steering and proceeding toward the rear axle. The last paragraphs are devoted to the power take-off and power lift systems. Interspersed where needed are additional tabular specifications pertaining to wear limits, torquing, etc.

HOW TO USE THE INDEX

Suppose you want to know the procedure for R&R (remove and reinstall) of the engine camshaft. Your first step is to look in the index under the main heading of ENGINE until you find the entry "Camshaft." Now read to the right where under the column covering the tractor you are repairing, you will find a number which indicates the beginning paragraph pertaining to the camshaft. To locate this wanted paragraph in the manual, turn the pages until the running index appearing on the top outside corner of each page contains the number you are seeking. In this paragraph you will find the information concerning the removal of the camshaft.

More information available at Clymer.com
Phone: 805-498-6703

Haynes Publishing Group
Sparkford Nr Yeovil
Somerset BA22 7JJ England

Haynes North America, Inc
859 Lawrence Drive
Newbury Park
California 91320 USA

ISBN-10: 0-87288-416-3
ISBN-13: 978-0-87288-416-8

© **Haynes North America, Inc. 1990**
With permission from J.H. Haynes & Co. Ltd.

Clymer is a registered trademark of Haynes North America, Inc.

Printed in Malaysia.

Cover art by Sean Keenan

SHOP MANUAL
CASE/INTERNATIONAL

MODELS
385—485—585—685—885

Engine serial number is stamped on right side of engine crankcase on all models. Tractor model and serial number are stamped on a plate attached to right side of front axle support. Transmission serial number is stamped on a plate attached to the right rear side of rear main frame. Model and serial number plate for front drive axle is located on left rear of axle housing.

INDEX (By Starting Paragraph)

INDEX (CONT.)

INDEX (CONT.)

INDEX (CONT.)

DUAL DIMENSIONS

This service manual provides specifications in both the U.S. Customary and Metric (SI) system of measurements. The first specification is given in the measuring system used during manufacture, while the second specification (given in parenthesis) is the converted measurement. (For instance, a specification of "0.011 inch (0.279 mm)" would indicate that the equipment was manufactured using the U.S. system of measurement and the metric equivalent of 0.011 inch is 0.279 mm.

CONDENSED SERVICE DATA

	385 Diesel	485 Diesel	585 Diesel	685 Diesel	885 Diesel
GENERAL					
Engine Make			Case/International		
Engine Model	D-155	D-179	D-206	D-239	D-268
Number of Cylinders	3	3	4	4	4
Bore	3.875 in. (98.4 mm)	3.875 in. (98.4 mm)	3.875 in. (98.4 mm)	3.875 in. (98.4 mm)	3.937 in. (100.0 mm)
Stroke	4.37 in. (111.1 mm)	5.06 in. (128.5 mm)	4.37 in. (111.1 mm)	5.06 in. (128.5 mm)	5.51 in. (140.0 mm)
Displacement	155 cu. in. (2.54 L)	179 cu. in. (2.93 L)	206 cu. in. (3.38 L)	239 cu. in. (3.92 L)	268 cu. in. (4.39 L)
Main Bearings, Number of	4	4	5	5	5
Cylinder Sleeves	Wet	Wet	Wet	Wet	Wet
Forward Speeds, Number of	8 (1)	8 (1)	8 (1)	8 (1)	8 (1)
Alternator/Starter Make . . .			Lucas and Bosch		
Battery Ground			Negative		

(1) Sixteen forward speeds, if equipped with Two-Speed Power Shift.

CONDENSED SERVICE DATA (CONT.)

	385 Diesel	485 Diesel	585 Diesel	685 Diesel	885 Diesel
TUNE-UP					
Compression Pressure.....			315-340 psi (2) (2172-2344 kPa)		
Firing Order..........	1-3-2	1-3-2		1-3-4-2	
Valve Tappet Gap (Hot), Intake and Exhaust.....			0.012 in. (0.304 mm)		
Valve Face Angle........			45°		
Valve Seat Angle........			44°		
Injection Pump Make.....			Robert Bosch		
Injection Pump Timing....			See Paragraph 61.		
Engine Low Idle — Rpm..			725-750		
Engine High Idle — Rpm (No Load).............	2340	2480	2590	2705	2650
Engine Full Load — Rpm..	2200	2180	2300	2300	2400

(2) Approximate psi (kPa), at sea level, at cranking speed.

SIZES-CLEARANCES-CAPACITIES

	385 Diesel	485 Diesel	585 Diesel	685 Diesel	885 Diesel
Crankshaft Main Journal Diameter.............			3.1484-3.1492 in. (79.97-79.99 mm)		
Crankpin Diameter.......			2.5185-2.5193 in. (63.97-63.99 mm)		
Camshaft Journal Diameter			2.2823-2.2835 in. (57.97-58.00 mm)		
Piston Pin Diameter......			1.4172-1.4174 in. (35.997-36.002 mm)		
Valve Stem Diameter, Intake...............			0.3920-0.3924 in. (9.957-9.967 mm)		
Exhaust.............			0.3912-0.3915 in. (9.936-9.945 mm)		
Main Bearing Diametral Clearance............			0.0028-0.0055 in. (0.071-0.140 mm)		
Rod Bearing Diametral Clearance............			0.002-0.004 in. (0.051-0.102 mm)		
Piston Skirt Diametral Clearance...........			0.0035-0.0050 in. (0.09-0.13 mm)		
Crankshaft End Play......			0.006-0.009 in. (0.152-0.229 mm)		
Camshaft Bearing Diametral Clearance....			0.0009-0.0033 in. (0.023-0.084 mm)		
Camshaft End Play.......			0.004-0.018 in. (0.10-0.45 mm)		
Cooling System Capacity..	3.0 U.S. gal. (11.4 L)			3.6 U.S. gal. (13.7L)	
Crankcase Oil (With Filter)	8 U.S. qts. (7.5 L)			10 U.S. qts. (9.5 L)	

CONDENSED SERVICE DATA (CONT.)

	385 Diesel	485 Diesel	585 Diesel	685 Diesel	885 Diesel

SIZES-CLEARANCES-CAPACITIES (CONT.)

Transmission, Differential
and Hydraulics, Capacity _____ 36 U.S. qts. (3)_____
(34 L)

Oil Type. . . . . _____ Hy-Tran Plus _____

Front Drive Axle Capacity. _____ 4.75 U.S. qts. _____
(4.5 L)

Oil Type. . . . . _____ SAE 90 EP _____

Front Drive Axle Hub
(Each Side) . . . . _____ 0.8 U.S. qt. _____
(0.75 L)

Oil Type. . . . . _____ SAE 90 EP _____

(3) If equipped with front drive axle, add 2.6 U.S. qts. (2.5 L).

FRONT AXLE (TWO-WHEEL DRIVE)

FRONT WHEEL BEARING

All Models

1. Refer to Figs. 1 and 2 for typical wheel hub and bearing assemblies.

The tapered inner and outer roller bearings are not interchangeable. Clean and inspect bearing cups (3 and 5) and cones (2 and 6) and renew as necessary.

NOTE: On adjustable axle hubs, if wear ring (12) is excessively worn, renew wear ring.

Install inner seal (1) on spindle and coat lips of seal with grease. Pack bearings and fill hub cavity with No. 2 lithium grease. Install inner bearing cone (2). Install hub assembly (4), outer bearing cone (6), washer (7) and nut (9). Adjust wheel bearing preload as follows: Tighten nut to a torque of 70 ft.-lbs. (95 N·m) while rotating the hub. Then, loosen nut and retorque to 50 ft.-lbs. (68 N·m) on Models 385, 485 and 585 with standard axles or 25 ft.-lbs. (34 N·m) on Models 385, 485 and 585 equipped with heavy duty axles and all Model 685 and 885 tractors. Then,

loosen nut 1/4 turn to align slot in nut with hole in spindle and install cotter pin (8). Install lithium grease in end of hub and in cap (10) and install cap.

SPINDLES

All Models With Adjustable Axles

2. R&R SPINDLES. To remove either spindle (9—Figs. 3, 4 or 5), block rear wheels securely and apply park brake. Raise front of tractor and install jack stand under axle main member (13). Unbolt and remove front wheel and hub. Disconnect tie rod from steering arm (2 or 6). Remove bolt (1) and nut (3) and lift off steering arm. Remove Woodruff key (11) and lower spindle (9) with thrust bearing (8) and felt ring (10), downward out of axle extension. Remove thrust bearing and felt ring from spindle.

With spindle removed, check spindle bushings (7) in axle extension for excessive wear or other damage and renew as required. Install new bushings (7) so that lubrication grooves are toward the grease fitting. Drive bushings in from top and bottom of extension until each bushing is 0.060 inch (1.5 mm) below the surface.

When reassembling, install new felt ring (10) and thrust bearing (8) as necessary.

NOTE: Thrust bearing (8) is marked TOP for correct installation.

Install spindle assembly, Woodruff key and steering arm. Tighten steering arm clamping bolt to a

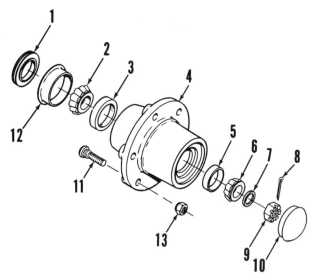

Fig. 1—Exploded view of front wheel hub and bearings used on all two-wheel drive models equipped with adjustable front axles.

1. Oil seal	
2. Bearing cone (inner)	
3. Bearing cup (inner)	7. Washer
	8. Cotter pin
4. Hub	9. Castellated nut
5. Bearing cup (outer)	10. Cap
	11. Lug bolt
6. Bearing cone (outer)	12. Wear ring
	13. Lug nut

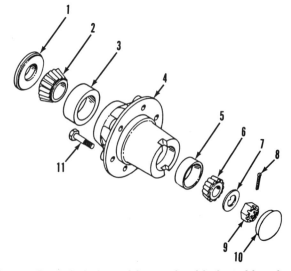

Fig. 2—Exploded view of front wheel hub and bearings used on two-wheel drive models equipped with nonadjustable (cast) front axle. Refer to Fig. 1 for legend.

torque of 80-90 ft.-lbs. (108-122 N·m). Install tie rod and tighten nut to a torque of 50 ft.-lbs. (68 N·m).

Check oil seal (1—Fig. 1) and renew as necessary as outlined in paragraph 1. Install hub and adjust bearings as in paragraph 1.

All Models With Nonadjustable Cast Axle

3. R&R SPINDLES. To remove either spindle (11—Fig. 6), block rear wheels securely and apply park brake. Raise front of tractor and install jack stand under axle main member (16). Unbolt and remove front wheel and hub.

Remove lubrication fitting from steering arm (5 or 10). Remove three bolts (22) and nuts, then move steering arm, steering cylinder (left side) and tie rod

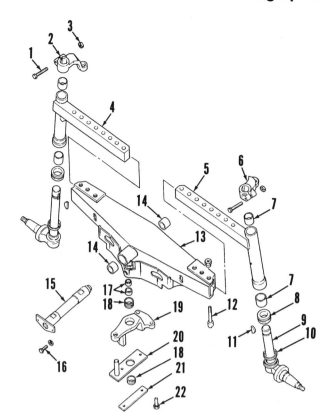

Fig. 4—Exploded view of heavy duty straight adjustable front axle used on some tractors. Refer to Fig. 3 for legend.

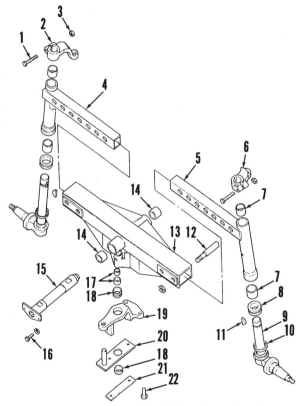

Fig. 3—Exploded view of straight adjustable front axle used on some tractors.

1. Bolt	
2. Steering arm	12. Shoulder bolt
(R.H.)	13. Axle main
3. Nut	member
4. Axle extension	14. Bushings
(R.H.)	15. Pivot pin
5. Axle extension	16. Cap screw (2)
(L.H.)	17. Steel bushings
6. Steering arm	18. Bushings
(L.H.)	19. Steering arm
7. Bushings	(center)
8. Thrust bearing	20. Steering cylinder
9. Spindle	retaining bar
10. Felt ring	21. Lockplate
11. Woodruff key	22. Cap screw (2)

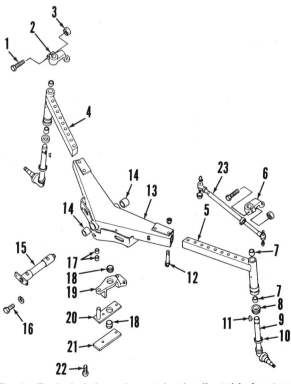

Fig. 5—Exploded view of swept back adjustable front axle used on some tractors. Refer to Fig. 1 for legend except for tie rod assembly (23).

rearward from axle. Remove cap screws (1), caps (2) and gaskets (3). Remove nut and lockwasher from lock pin (15). Drive lock pin forward out of axle main member. Using a hammer and punch, drive kingpin (12) out of axle and spindle. Remove spindle, thrust bearing (14) and shim (13) if so equipped.

Inspect kingpin bushings (4) and renew if necessary. Bushings should be installed until inner ends of bushings are 0.016-0.031 inch (0.4-0.8 mm) below the inside surfaces of spindle bores. Lubrication grooves in bushings must align with lubrication fittings.

Install a new thrust bearing (14), with TOP mark up, on bottom inside edge of spindle. Install spindle over end of axle main member with thrust bearing under axle. Use a jack to hold spindle and bearing tight against bottom of axle. Using a feeler gage, measure gap between upper edge of axle and the spindle. If gap is more than 0.005 inch (0.127 mm), install shims to reduce gap to less than 0.005 inch (0.127 mm). Shims are available in thicknesses of 0.005 and 0.010 inch (0.127 and 0.254 mm).

Inspect kingpin (12) and renew as necessary. Install kingpin with flat nearest to the top and in alignment with hole for the lock pin (15). Install lock pin, lockwasher and nut. Tighten nut securely. Install steering arm and tighten bolts (22) and nuts to a torque of 160-180 ft.-lbs. (217-244 N·m). Install lubrication fitting in steering arm.

Check oil seal (1—Fig. 2) and install new seal, if needed, then install hub and adjust bearings as outlined in paragraph 1.

AXLE MAIN MEMBER AND PIVOT PIN

All Models With Adjustable Axles

4. REMOVE AND REINSTALL. To remove axle main member (13—Figs. 3, 4 and 5), first block rear wheels securely and apply park brake. If so equipped, remove front weights and weight bracket. Unbolt and remove front cover. Support tractor with stands under side rails or oil pan, and remove front wheels. Disconnect tie rod ends from left and right steering arms (2 and 6). Unbolt and remove axle extensions (4 and 5) with spindle assemblies. Identify and disconnect steering cylinder hoses at cylinder. Cap and plug all openings. Remove the cylinder anchor clevis pin. Straighten corners of lock plate (21) and remove cap screws (22), lock plate (21) and cylinder retaining bar (20). Hold cylinder upward and slide cylinder clevis from anchor on axle. Remove steering cylinder. Remove center steering arm (19) and tie rods.

Place a floor jack under axle main member (13). Remove cap screws (16) and thread a slide hammer puller into front end of pivot pin (15). Remove pivot pin, then lower axle main member and remove from under tractor.

Inspect axle pivot bushings (14) and renew if necessary, aligning lubrication holes with lubrication fittings. Inspect cylinder pivot bushings (18) and steering arm pivot bushings (17) and renew as necessary.

Reinstall by reversing removal procedure, keeping the following points in mind: Tighten cap screws (16) and (22) to a torque of 80-90 ft.-lbs. (110-124 N·m). Bend corners of lock plate (21) against flat of bolt head (22). Tighten nuts on shoulder bolts (12) to a torque of 246-272 ft.-lbs. (334-369 N·m). Tighten tie rod end slotted nuts to a torque of 50 ft.-lbs. (68 N·m).

All Models With Nonadjustable Cast Axle

5. REMOVE AND REINSTALL. To remove axle main member (16—Fig. 6), block rear wheels securely and apply park brake. If so equipped, remove front weights and weight bracket. Unbolt and remove front cover. Raise front of tractor and support with stands under side rails or oil pan. Remove front wheels. Identify and disconnect steering cylinder hoses at cylinder. Cap or plug all openings immediately to pre-

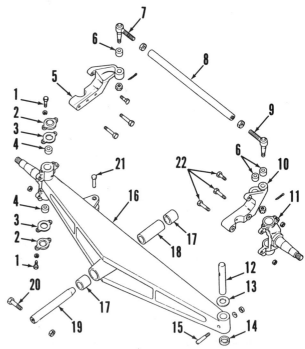

Fig. 6—Exploded view of heavy duty (cast) nonadjustable front axle used on some tractors.

1. Cap screw
2. Cap
3. Gasket
4. Bushing
5. Steering arm (R.H.)
6. Dust seal
7. Tie rod end (R.H.)
8. Tie rod tube
9. Tie rod end (L.H.)
10. Steering arm (L.H.)
11. Spindle
12. King pin
13. Shim
14. Thrust bearing
15. King pin lock pin
16. Axle main member
17. Bushings
18. Spacer
19. Pivot pin
20. Bolt
21. Cylinder pin
22. Bolts

vent dirt from entering system. Place a floor jack under axle main member to support axle assembly. Remove clamp bolt (20) and nut, then thread a slide hammer puller into end of pivot pin (19). Remove pivot pin, then lower axle main member assembly and remove from under tractor.

Remove tie rod and steering cylinder. Remove spindles if desired as outlined in paragraph 3. Inspect bushings (17) and renew if necessary. Install new bushings (17) and spacer (18) so that bushings are slightly below flush with outer edges of bore in axle main member.

Reinstall by reversing removal procedure keeping the following points in mind: When installing pivot pin (19), flat for clamping bolts (20) must fact downward. Tighten clamping bolt and nut to a torque of 80-90 ft.-lbs. (108-122 N·m). Tighten tie rod end slotted nuts to a torque of 50 ft.-lbs. (68 N·m). Lubricate pivot bushings with No. 2 lithium grease.

TIE RODS AND TOE-IN

All Models (Two-Wheel Drive)

6. Two types of tie rods used with adjustable axles are shown in Fig. 7. Two tie rods are used on each tractor. Refer to Fig. 6 for tie rod used on models equipped with the nonadjustable cast axle. Removal and disassembly of tie rods on all models is obvious after examination of unit and reference to Figs. 6 and 7.

Front wheel toe-in on all models equipped with adjustable axles should be 3/16-5/16 inch (5-8 mm). To adjust toe-in, remove clamp bolt (2—Fig. 7) or set screw (2A) and loosen jam nut (4). Rotate tube (3)

to lengthen or shorten tie rod as required. Adjust both tie rods equally. Reassemble tie rods and recheck toe-in. Readjust if necessary.

Front wheel toe-in on all models equipped with the nonadjustable cast axle should be 0-5/32 inch (0-4 mm). To adjust toe-in, refer to Fig. 6 and loosen jam nuts at each end of tube (8). One nut is left hand thread and the other is right hand thread. Rotate tube as required to lengthen or shorten tie rod assembly. Tighten jam nuts and recheck toe-in. Readjust if necessary.

FRONT SUPPORT

All Models (Two-Wheel Drive)

7. REMOVE AND REINSTALL. To remove the front support (3—Fig. 8 or 9), first block rear wheels securely and apply park brake. Remove hood, grille and side panels. Drain cooling system. Remove radi-

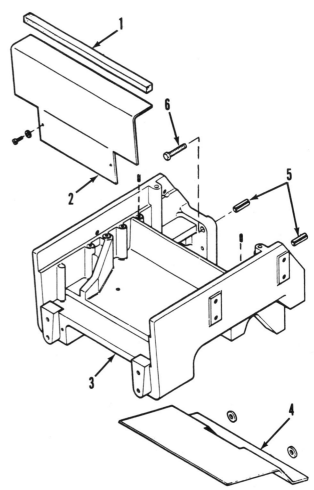

Fig. 8—View of cast front support used on some two-wheel drive models.

1. Foam seal	4. Lower air baffle
2. Front cover	5. Spring pin dowels
3. Front support	6. Cap screws

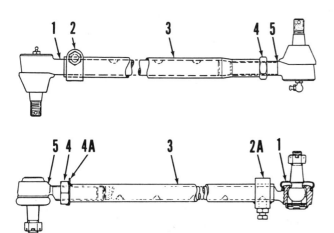

Fig. 7—View showing typical tie rod assemblies used on models equipped with adjustable axles.

1. Tie rod end (outer)	3. Tube
2. Clamp & bolt	4. Jam nut
2A. Collar & set screw	4A. Lockwasher
	5. Tie rod end (inner)

ator as outlined in paragraph 68. Remove front axle as outlined in paragraph 4 or 5.

On models with cast front support, disconnect upper steering lines. Cap or plug all openings. On all models, unbolt and remove oil pan shield. Attach a hoist to front support, then unbolt and remove front support (cast support) or front support and side plate assembly (fabricated support).

Reinstall by reversing removal procedure. Tighten front support mounting 5/8 inch bolts to 220 ft.-lbs. (298 N·m) torque and 3/4 inch bolts to 400 ft.-lbs. (542 N·m) torque.

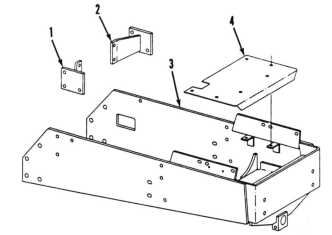

1. Engine mount (right)
2. Engine mount (left)
3. Front support
4. Air baffle plate

Fig. 9—View of fabricated front support used on some two-wheel drive models.

FRONT-WHEEL DRIVE

All models are available with a ZF mechanical front drive axle assembly. Axle Models APL-324 and APL-330 are used. Refer to Model and Serial Number plate on left rear of axle housing for identification.

DRIVE AXLE ASSEMBLY AND FRONT SUPPORT

All Models So Equipped

8. REMOVE AND REINSTALL. To remove the front drive axle assembly, block rear wheels securely and apply park brake. Unbolt and remove drive shaft shield and front drive shaft assembly. Disconnect steering cylinder lines and plug or cap all openings. If so equipped, unbolt and remove front weights and weight bracket. Loosen front wheel to hub nuts on both wheels. Raise front of tractor until front wheels are off the floor. Support front of tractor with splitting tool MS2700B under engine oil pan. Remove wheel nuts and using a hoist, remove front wheels. Place a floor jack under axle assembly. Unbolt pivot bearing caps (9 and 10—Fig. 10), lower axle assembly and remove from under tractor.

Clean and inspect bearings (8) for excessive wear and renew as necessary. Renew "O" rings (7) and reinstall axle assembly by reversing removal procedure. Tighten cap screws (11 and 13) to a torque of 133 ft.-lbs. (180 N·m).

9. R&R FRONT SUPPORT. To remove front support (3—Fig. 10), first remove hood, grille and side panels. Drain cooling system and remove radiator as outlined in paragraph 68. Block rear wheels securely and apply park brake. Remove front drive axle assembly as outlined in paragraph 8. Disconnect upper front steering lines as necessary. Cap or plug all openings. Attach a hoist to front support, then unbolt and remove front support.

Reinstall by reversing removal procedure. Tighten front support mounting 5/8 inch cap screws to a torque of 220 ft.-lbs. (298 N·m).

WHEEL HUB AND PLANETARY

All Models So Equipped

10. R&R AND OVERHAUL. To remove the wheel hub and planetary, support axle housing and remove wheel and tire assembly.

NOTE: Although most work on the front drive axle assembly can be accomplished without removing the complete axle assembly, some mechanics prefer to remove the assembly if total overhaul of front axle is to be performed.

Rotate wheel hub until drain plug (17—Fig. 11) is at bottom, remove plug and drain oil. Remove the two socket head screws (16), then remove planetary carrier (15). Remove snap ring (22) from each of the planetary gears. Using a suitable puller, remove the planetary gears. Remove angle ring (19) and roller bearing (21) from each planetary gear (20).

Remove snap ring (13) and thrust washer (11) from stub axle (sun gear) (12). Remove eight cap screws (23), then screw four M10 × 1.5 mm, 60 mm long cap screws into every other hole. Make certain that all four bolt heads are equal distance from inner face of ring gear (10). Using a suitable three legged puller and thrust piece pushing against the bolt heads, remove ring gear. Then, using the same puller and thrust piece, remove hub (6) and outer bearing cone (4). Inspect hollow dowels (9) for wear and if worn or otherwise damaged, thread inside of dowels. Use a screw and slide hammer puller to remove dowels.

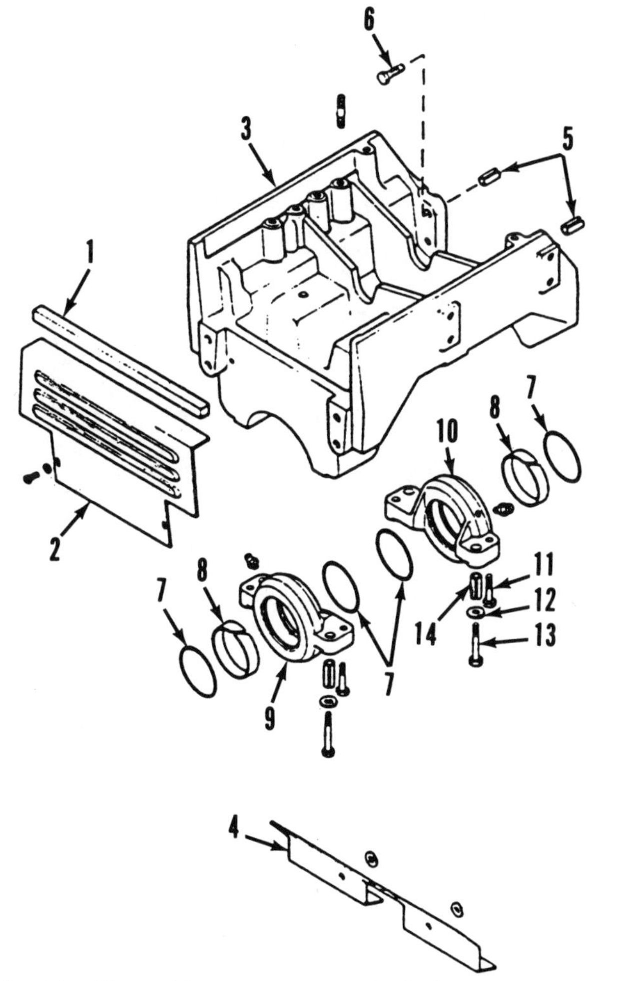

Remove and discard "O" ring (8). Remove scraper ring (2) and oil seal (3) from hub (6), then drive cups for bearings (4) from hub. Using a puller, remove inner bearing cone (4) from swivel housing (32). Check

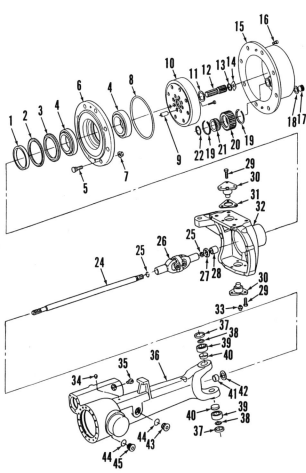

Fig. 11—Exploded view of wheel hub and planetary, swivel housing and axle shaft and left axle housing and relative components. Except for axle housing, right side is similar.

1. Wear ring
2. Scraper ring
3. Oil seal
4. Bearing assemblies
5. Lug bolt
6. Hub
7. Lug nut
8. "O" ring
9. Hollow dowel pin
10. Ring gear
11. Thrust washer
12. Stub shaft (sun gear)
13. Snap ring
14. Shim
15. Planetary carrier
16. Socket head screw
17. Drain/level plug
18. Seal ring
19. Angle rings
20. Planetary gear
21. Roller bearing
22. Snap ring
23. Cap screw
24. Axle shaft
25. Snap rings
26. Double "U" joint
27. Oil seal
28. Sleeve bushing
29. Cap screw
30. King pins
31. Shims (upper & lower)
32. Swivel housing
33. Lube fitting
34. Breather vent
35. Air bleed screw
36. Axle housing (L.H.)
37. Cap ring
38. "O" ring
39. Bearing assy.
40. Cup plug
41. Sleeve bushing
42. Oil seal
43. Level plug
44. "O" rings
45. Drain plug

Fig. 10—View of front support and relative components used on four-wheel drive models.

1. Foam seal
2. Front cover
3. Front support
4. Lower air baffle
5. Spring pin dowels
6. Cap screws
7. "O" rings
8. Pivot bearings
9. Bearing cap (front)
10. Bearing cap (rear)
11. Cap screw
12. Washer
13. Cap screw
14. Spring pin dowel

wear ring (1) and if worn, use a chisel to cut the ring. Do not cut the surface of swivel housing. If necessary to remove stub shaft (sun gear) (12), remove swivel housing (32) as outlined in paragraph 11. Remove thrust shim (14) from center of planetary carrier (15). Clean and inspect all parts and renew any showing excessive wear or other damage.

NOTE: Wheel hub bearings (4) are available only in sets of two and must be installed in sets. There is no adjustment for the bearings and hub rolling torque depends upon condition of the bearings.

When reassembling, heat new wear ring (1) and push ring into position on swivel housing. Heat inner bearing cone (4) to a temperature of 250° F (120° C) in a bearing oven, then install on swivel housing against shoulder.

WARNING: Always wear heat protective gloves when handling heated parts.

Install inner and outer bearing cups (4) until seated in hub. Apply a light coat of sealing compound to outer edge of seal (3) and scraper ring (2), then install seals with lip facing inward. Install hub assembly on swivel housing. Heat outer bearing cone (4) to a temperature of 250° F (120° C) in a bearing oven and install the cone on swivel housing firmly against hub bearing cup. When bearing has cooled, hub should have zero end play. Drive hollow dowel pins (9) into swivel housing (32), then drive ring gear (10) over the hollow dowels on swivel housing. Apply Loctite stud lock to threads of cap screws (23), install and tighten cap screws to a torque of 67 ft.-lbs. (90 N·m). Install ''O'' ring (8) in groove on hub (6). Install thrust washer (11) and snap ring (13).

Lubricate roller bearing (21), install in planetary gear (20) and secure with angle rings (19). Heat planetary gear assembly and install in planetary carrier (15). Repeat the procedure and install the other two planetary gear assemblies. Secure with snap rings (22).

To determine correct thickness of axle shaft end play shim (14), proceed as follows: Refer to Fig. 12 and measure distance from planetary carrier flange to recess for end play shim. Record this measurement. Hold axle shaft and sun gear inward, refer to Fig. 13 and measure distance from end of stub shaft (sun gear) to hub outer flange. Subtract last measurement from first measurement to obtain full axle shaft end play. Select a shim which will provide end play of 0.012-0.024 inch (0.3-0.6 mm). Shims are available in thicknesses of 0.039, 0.059, 0.067, 0.079, 0.087 and 0.098 inch (1.0, 1.5, 1.7, 2.0, 2.2 and 2.5 mm). Apply Loctite stud lock to one side of selected shim and stick shim in place as shown in Fig. 14. Use a hammer and center punch to peen metal (2) to hold shim in place. Install planetary carrier (15—Fig. 11) over

lug bolts (5) and install the socket head screws (16). Tighten to a torque of 70 ft.-lbs. (96 N·m). Rotate hub and planetary until level plug (17) is horizontal to center of hub. Fill hub with 0.8 U.S. quart (0.75 L) of SAE 90 EP gear oil. Tighten plug with new seal ring (18) securely.

Fig. 12—Measure distance from planetary carrier flange to recess for end play shim.

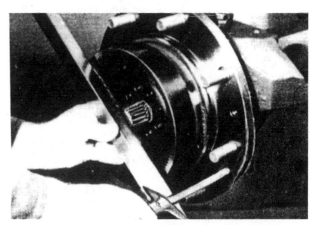

Fig. 13—Measure distance from end of stub shaft (sun gear) to hub outer flange.

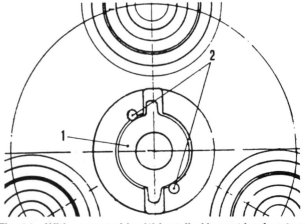

Fig. 14—With correct shim (1) installed in seat in planetary carrier, use a center punch to peen metal (2) to hold shim in place.

Install wheel and tire assembly and tighten lug nuts (7) to a torque of 222-259 ft.-lbs. (300-350 N·m). Lower tractor to the floor.

Repeat operation on opposite side as required.

SWIVEL HOUSING AND AXLE SHAFT

All Models So Equipped

11. R&R AND OVERHAUL. To remove the swivel housing and axle shaft, first remove the wheel hub and planetary as outlined in paragraph 10. Then, refer to Fig. 11 and proceed as follows: Remove drain plug (45) and drain oil. Disconnect tie rod end from swivel housing. Remove six cap screws (29) and withdraw king pins (30) and shims (31) which are used on both king pins. Lift off swivel housing. Remove oil seal (27) and sleeve bushing (28) from swivel housing.

Withdraw axle shaft assembly. Remove snap rings (25) and separate axle shaft (24) and stub shaft (sun gear) (12) from double "U" joint (26). Remove cap ring (37), "O" ring (38), bearing assembly (39) and cup plug (40) from top and bottom of axle housing. Remove oil seal (42) and sleeve bushing (41) from end of axle housing.

Clean and inspect all parts and renew as necessary. When reassembling, drive new bushing (41) and oil seal (42) into end of axle housing. Lip of seal must face sleeve bushing. "U" joint (26) is available only as an assembly. Reassemble axle shaft and stub shaft (sun gear) to the double "U" joint by reversing the disassembly procedure. Apply a light coat of sealing compound to outer edges of cup plugs (40) and drive cup plugs into axle housing. Install bearing cups (39) until seated. Lubricate lips of oil seal (42) and carefully insert axle shaft into axle housing. Make certain axle shaft splines engage splines in differential side gear. Install new sleeve bushing (28) and oil seal (27) in swivel housing (32). Lip of seal must face bushing. Lubricate bearing cones (39) and cap rings (37) and install in axle housing. Apply grease to new "O" rings (38) and install on king pins (30). Lubricate lips of oil seal (27), then carefully install swivel housing (32) over stub shaft (sun gear) (12) and axle housing. Install lower king pin and shim assembly and tighten cap screws finger tight. Install upper king pin with original shim (31) and cap screws finger tight. Torque the bottom cap screws (29) first, then the upper cap screws to 89 ft.-lbs. (120 N·m). Check swing resistance with a torque wrench which should be 6-8 ft.-lbs. (8-11 N·m). Adjust swing resistance by using alternate shim thickness. Shims (31) are available in thicknesses of 0.008, 0.012, 0.016, 0.024, 0.031, 0.039, 0.055, 0.063, 0.071 and 0.079 inch (0.2, 0.3, 0.4, 0.6, 0.8, 1.0, 1.4, 1.6, 1.8 and 2.0 mm). Divide total shim thickness and install equally at top and bottom.

Reinstall wheel hub and planetary as outlined in paragraph 10. Connect tie rod end to swivel housing arm and tighten nut to a torque of 74-81 ft.-lbs. (100-110 N·m).

Remove level plug (43) and fill axle housing to level plug opening with SAE 90 EP gear oil. Refill capacity is 4.75 U.S. quarts (4.5 L).

DIFFERENTIAL AND BEVEL DRIVE GEARS

All Models So Equipped

12. R&R AND OVERHAUL. To remove the front differential assembly, securely block rear wheels and apply park brake.

NOTE: Although most work on the front drive axle assembly can be accomplished without removing the complete axle assembly, some mechanics prefer to remove the axle assembly when overhauling the differential and bevel drive pinion.

Unbolt and remove drive shaft shield and front drive shaft assembly as outlined in paragraph 16. Disconnect steering cylinder lines and cap or plug all openings. If so equipped, unbolt and remove front weights and weight bracket. Raise front of tractor until front wheels are off the floor. Support tractor under oil pan using tractor splitting tool MS2700B or equivalent. Support each end of axle housing with jack stands. Unbolt and remove front wheels. Disconnect tie rod ends from arms on swivel housings. Remove drain plug (45—Fig. 11) and drain oil from axle housing. Attach a sling and hoist to planetary hub. Unbolt and remove upper and lower king pins (30) with bearing cones (39). Keep shims (31) with respective king pins. Remove planetary, swivel housing and axle shaft from each end of axle housing. Place a floor jack under center of axle housing. Unbolt axle pivot bearing caps (9 and 10—Fig. 10), lower axle assembly and remove from under tractor. Unbolt and remove right axle housing (23—Fig. 15) from left axle housing (5). Then, remove differential assembly from left axle housing.

NOTE: Two types of differential units are used. Type One (without limited slip clutch) is shown in Fig. 15. Type Two shown in Fig. 16 is equipped with limited slip self-locking clutch discs.

13. To disassemble Type One differential, refer to Fig. 15 and using a suitable puller remove bearing cones (1 and 26). Pry off lock plate (10) and remove cap screws (11). Remove ring gear (12) from differential case (17). Remove roll pin (19) and pinion shaft (18), then separate thrust washers (13), differential pinions (15) and side gears (14 and 16) from case. Remove roll pins (20 and 21). Remove bearing cups (1 and 26) and shim spacers (2 and 25) from axle housings (5 and 23). Clean and inspect all parts for excessive wear or other damage and renew as necessary. Ring gear (12) and bevel drive pinion (8—Fig. 19) are available only as a matched set. Refer to paragraph 15 for R&R of bevel drive pinion. When reassembling Type One differential, refer to Fig. 15 and install thrust washers (13), side gears (14 and 16) and differential pinions (15) in case (17). Install pinion shaft into case and through differential pinions (15). Secure shaft with roll pin (19). Screw two guide studs into case (17) and press ring gear (12) onto the case.

Install roll pins (20 and 21) and cap screws (11). Tighten cap screws to a torque of 51 ft.-lbs. (69 N·m), then press lock plate (10) over bolt heads. Heat bearing cones (1 and 26) in a bearing oven to a temperature of 250° F (120° C) and install on differential assembly.

WARNING: Always wear heat protective gloves when handling heated parts.

14. To disassemble the Type Two differential, use a suitable puller and remove carrier bearing cones. Pry off lock plate (10—Fig. 16) and remove cap screws (11). Separate ring gear (12) from differential case (17). Remove roll pin (19) and pinion shaft (18), then remove differential pinions (15), thrust washers (13), side gears (14 and 16) with thrust discs (1), friction plates (2) and friction discs (3). Remove roll pins (20 and 21). Remove bearing cups (1 and 26—Fig. 15) and shim spacers (2 and 25) from axle housings (4 and 23). Clean and inspect all parts for excessive wear or other damage and renew as necessary. Ring gear (12—Fig. 16) and bevel drive pinion (8—Fig. 19) are available only as a matched set. Refer to paragraph 15 for R&R of bevel drive pinion. When reassembling, install one

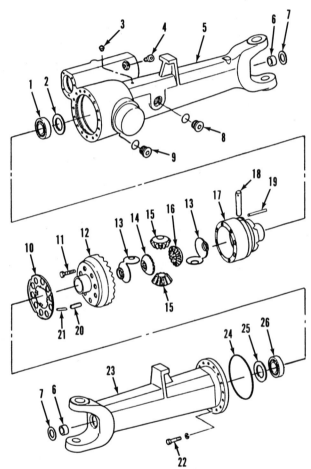

Fig. 15—View of axle housings and exploded view of differential (without limited slip). Refer also to Fig. 16.

1. Carrier bearing assy.	
2. Shim washer	15. Differential pinions
3. Breather vent	16. Side gear (L.H.)
4. Air bleed screw	17. Differential case
5. Axle housing (L.H.)	18. Pinion shaft
6. Sleeve bushing	19. Roll pin
7. Oil seal	20. Roll pin (large)
8. Oil level plug	21. Roll pin (small)
9. Drain plug	22. Cap screw
10. Lock plate	23. Axle housing (R.H.)
11. Cap screw	24. "O" ring
12. Ring gear	25. Shim spacer
13. Thrust washers	26. Carrier bearing assy.
14. Side gear (R.H.)	

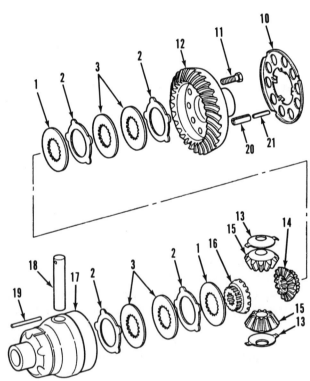

Fig. 16—Exploded view of limited slip differential assembly.

1. Thrust disc	15. Differential pinions
2. Friction plates	16. Side gear (L.H.)
3. Friction discs	17. Differential case
4.	18. Pinion shaft
10. Lock plate	19. Roll pin
11. Cap screws	20. Roll pin (large)
12. Ring gear	21. Roll pin (small)
13. Thrust washer	
14. Side gear (R.H.)	

0.118 inch (3.0 mm) thick thrust disc (1—Fig. 16) on each side gear, then starting with a friction plate (2) alternate with friction disc (3) until five friction plates and four friction discs are in place on each side gear. Install side gear (16) assembly into differential case (17). Install thrust washers (13) and differential pinions (15). Install pinion shaft (18) through differential pinions and secure with roll pin (19). Then, install side gear (14) assembly in differential case. With differential case standing on carrier bearing end, apply light hand pressure on the plates and measure the distance between friction plate and end of case as shown in Fig. 17. Record measurement and check the clearance on opposite side, using a feeler gage through hole in case as shown in Fig. 18. Clearance (end play) of friction plates should be 0.0039-0.0078 inch (0.1-0.2 mm) for each side. If measurements are not correct, change thickness of thrust plate (1—Fig. 16) as needed. Thrust plates are available in thicknesses of 0.110, 0.114, 0.118, 0.122, 0.126 and 0.130 inch (2.8, 2.9, 3.0, 3.1, 3.2 and 3.3 mm). Screw two guide studs into case (17) and press ring gear (12) onto case. Install roll pins (20 and 21) and cap screws (11). Tighten cap screws to a torque of 51 ft.-lbs. (69 N·m), then press lock plate (10) over bolt heads. Heat carrier bearing cones in a bearing oven to a temperature of 250° F (120° C) and install on differential assembly.

WARNING: Always wear heat protective gloves when handling heated parts.

15. To remove the bevel drive pinion (8—Fig. 19), remove lock plate (1). Remove nut (2), then drive the bevel pinion shaft into left axle housing. Remove oil seal (3), outer bearing cone (4) and spacer (5). Drive shim spacer (6) and inner bearing cup (7) from housing, then remove outer bearing cup (4). Using a suitable puller, remove inner bearing cone (7) from bevel pinion shaft. Clean and inspect all parts and renew as necessary. Bevel drive pinion (8) and ring gear (12—Fig. 15 or 16) are available only as a matched set.

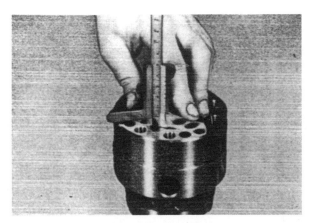

Fig. 17—When measuring friction plate clearance on ring gear side, measure distance from friction plate to end of differential case.

Before reinstalling bevel drive pinion, adjust differential carrier bearing preload as follows: Install a 0.47 inch (1.2 mm) thick shim spacer (2—Fig. 15) in left axle housing (5) and install left carrier bear-

Fig. 18—Using a feeler gage through hole in differential case, measure friction plate clearance on case side.

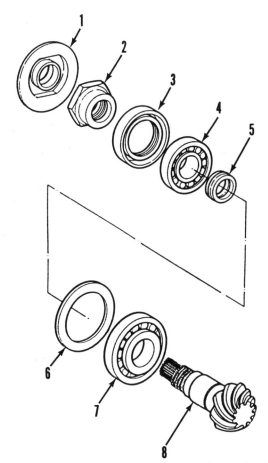

Fig. 19—Exploded view of bevel drive pinion and relative components.

1. Lock plate
2. Nut
3. Oil seal
4. Outer bearing assy.
5. Spacer ring
6. Shim spacer
7. Inner bearing assy.
8. Bevel drive pinion shaft

ing cup (1). Install differential assembly in left axle housing. Hold right carrier bearing cup on bearing cone and measure distance between outer edge of cup and left axle housing flange as shown in Fig. 20. Refer to Fig. 21 and measure distance between right axle housing flange and carrier bearing seat (shoulder). Subtract the first measurement from the second measurement. Select a shim washer (25—Fig. 15) with thickness equal to the difference between the two measurements. Shim washers are available in twenty thicknesses from 0.020-0.071 inch (0.5 to 1.8 mm). Install the selected shim washer (25) and carrier bearing cup in right axle housing. Remove differential assembly and wrap a cord around differential case. Attach a spring scale to loose end of cord. Reinstall differential assembly with spring scale protruding through bevel pinion shaft opening as shown in Fig. 22. Install right axle housing assembly and secure with four cap screws (22—Fig. 15) spaced at 90 degree intervals. Tighten the four cap screws to a torque of 44-52 ft.-lbs. (60-70 N·m). Pull on spring scale and check rolling resistance of differential assembly. Amount of pull on scale should be between 4.4 and 16.5 lbs. (2.0 and 7.5 kg). If not, remove right axle housing, remove carrier bearing

cup (26) and install a thicker shim washer (25) to increase or a thinner shim washer to decrease preload on bearings. Reassemble and reinstall right axle housing and recheck preload.

When preload is correct, measure distance "A" from end of bevel drive pinion bore to differential case as shown in Fig. 23. Record distance "A". Remove right axle housing and differential assembly. Install inner pinion shaft bearing cup. Hold pinion shaft bearing cone in position with a flat steel plate and measure distance "B" as shown in Fig. 24. Record distance "B". Measure diameter of differential case and divide by 2. This will be the case radius or distance ("D"—Fig. 25). Record distance "D". Distance "F" is the distance from inner bearing cone to the center of the differential case. Distance "A" plus distance "D" minus distance "B" will result in distance "F". Distance "F" minus the metric dimension "E" etched in millimeters on end of pinion shaft gear will give the metric shim spacer thickness to be installed at (S) between inner bearing cup (7—Fig. 19) and bearing cup seat (shoulder) in pinion housing. Shim spacers (6) are available in thicknesses of 0.035 to 0.059 inch (0.9 to 1.5 mm) in 0.002 inch (0.05 mm) increments.

Fig. 20—Measure distance between right carrier bearing cup and left axle housing flange.

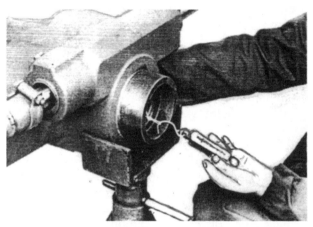

Fig. 22—Check carrier bearing preload with cord wrapped around differential case and connected to a spring scale.

Fig. 21—Measure distance between right axle housing flange and bearing cup seat (shoulder).

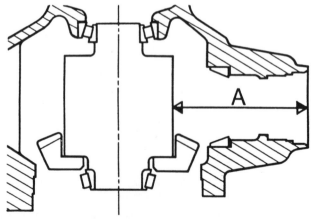

Fig. 23—Measure distance "A" between end of bevel drive pinion bore and differential case.

Using a bearing oven, heat inner bearing cone (7) to a temperature of 250° F (120° C) and install on bevel pinion shaft. Install outer bearing cup (4) in pinion housing until seated. Install special tool CAS-1881 measuring ring on pinion shaft so dowels are away from bearing. Install pinion shaft assembly into pinion housing. Heat bearing cone (4) in a bearing oven to a temperature of 250° F (120° C) and install on pinion shaft until seated in bearing cup. Install and tighten nut (2) on pinion shaft until a slight drag is felt when rotating bevel drive pinion. Wrap a cord around pinion shaft and attach a spring scale to end of cord. Check rolling resistance of pinion shaft and adjust by progressively tightening the nut. Rolling resistance must be 17.6-30.8 lbs. (8.0-14.0 kg) with new bearings or 11.0-17.6 lbs. (5.0-8.0 kg) with used bearings. Remove the pinion shaft and remove the measuring ring from pinion shaft. Using a micrometer, measure distance ("X"—Fig. 26). Select a spacer ring (5—Fig. 19) equal to the distance ("X"—Fig. 26). Spacer rings are available in fifteen thicknesses from 0.433 to 0.488 inch (11.0 to 12.4 mm) in increments of 0.0039 inch (0.1 mm). Install selected spacer ring on bevel pinion shaft and install the assembly in pinion housing. Heat outer bearing and install on pinion shaft until seated in bearing cup. Install and tighten pinion shaft nut to a torque of 384 ft.-lbs. (520 N·m). Again, check rolling resistance of the pinion shaft. If rolling resistance is not correct, disassemble and select a thicker spacer ring to increase or thinner spacer ring to decrease rolling resistance. Reinstall pinion shaft assembly.

Apply engineers blue to both sides of some of the ring gear teeth. Install differential and right axle housing. Install four cap screws (22—Fig. 15) at 90 degree intervals and tighten them to a torque of 44-52 ft.-lbs. (60-70 N·m). Rotate pinion shaft several times, then working through the drain plug opening, use a dial indicator and check ring gear to pinion backlash which should be 0.003-0.009 inch (0.07-0.22 mm). If backlash is less than 0.003 inch (0.07 mm), remove shims from right side and move them to left side. If backlash is more than 0.009 inch (0.22 mm), remove shims from left side and move them to right side. Unbolt and remove right axle housing and differential assembly. Check the tooth contact pattern on ring gear teeth and compare with patterns shown in Fig. 27. Views "A" and "B" shown correct tooth contact patterns. Views "C" and "D" show that pinion must be moved inward. Views "E" and "F" show that pinion must be moved outward. Readjust as necessary.

With ring gear and bevel drive pinion correctly adjusted, install differential assembly, lubricate and install "O" ring (24—Fig. 15) and install right axle housing (23). Install and tighten cap screws (22) to a torque of 44-52 ft.-lbs. (60-70 N·m).

Remove pinion nut (2—Fig. 19) and after applying a light coat of sealant to outer surface of oil seal (3), install oil seal with lip to inside. Install nut and tighten to a torque of 384 ft.-lbs. (520 N·m). Install new lock plate (1) over nut, then using a hammer and small punch, peen edge of plate into pinion shaft splines.

Install new oil seals (7—Fig. 15) if necessary, then reassemble and reinstall front drive axle by reversing disassembly and removal procedures. Fill axle housing to level plug opening with SAE 90 EP gear oil. Refill capacity is 4.75 U.S. quarts (4.5 L).

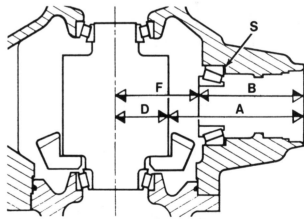

Fig. 25—Some of the measurements used in selecting correct shim spacer to be installed at "S".

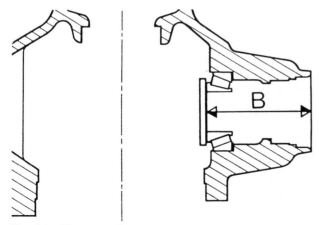

Fig. 24—Measure distance "B" between end of drive pinion shaft bore to steel plate as shown.

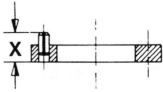

Fig. 26—Cutaway view of measuring ring CAS-1881. Use a micrometer to measure distance "X" from back of ring to tip of dowel pin.

FRONT DRIVE SHAFT AND SHIELD

All Models So Equipped

16. REMOVE AND REINSTALL. To remove the front drive shaft (2—Figs. 29 and 31), first remove drive shaft shield as follows: Block rear wheels securely and apply park brake. Refer to Fig. 28 and remove shield center bolts. Loosen shield rear bolts and place a support under shield (5). Remove shield front bolts, then remove the shield.

Remove front coupler cap screws (4—Fig. 29 and 31) and move front coupler rearward on drive shaft (2). Remove rear coupler cap screws (4) and pry rear coupler (5) forward on drive shaft. Support drive shaft assembly, remove support bar (1—Fig. 28) bolts and lower assembly from tractor. Remove nuts and washers and separate bearing retainer (4), shield support (3), shims (2) from support bar (1). If bearing (3—Figs. 29 and 31) is loose, noisy or rough, remove coupler (1 or 5) and remove bearing (3) from drive shaft (2). If bearing seat on drive shaft is worn, reverse ends of drive shaft when installing bearing.

When reassembling, install bearing on drive shaft in correct position and install the removed coupler so that bolt holes on both couplers align. Install bearing retainer (4—Fig. 28), shield support (3), shims (1) and support bar (1) on drive shaft bearing (3—Figs. 29 and 31). Do not tighten nuts at this time. Raise drive shaft assembly in place and tighten support bar (1—Fig. 28) bolts. Slide couplers on drive pinion shaft and transfer case output shaft. Align cap screw holes with grooves on shafts. Install coupler cap screws and tighten to a torque of 130 ft.-lbs. (175 N·m). Attach and zero a dial indicator against the bottom of drive shaft near the bearing. Tighten bearing retainer nuts and check drive shaft deflection on the dial indicator. Alignment is correct when a deflection of 0-0.002 inch (0-0.05 mm) is obtained. Change shims as necessary to obtain this deflection. Shims (2) are available in thicknesses of 0.020, 0.040 and 0.080 inch (0.5, 1.0 and 2.0 mm).

Reinstall shield by reversing the removal procedure.

TRANSFER GEARBOX

Two types of transfer gearboxes have been used. One type is the mechanical shift (dog clutch) shown in Fig. 29. The other type is equipped with an electric/hydraulic operated multiple disc clutch (Fig. 31). The clutch is engaged by three Belleville springs when hydraulic power is switched OFF. When hydraulic power is switched ON, the piston collapses the Belleville springs and clutch is disengaged.

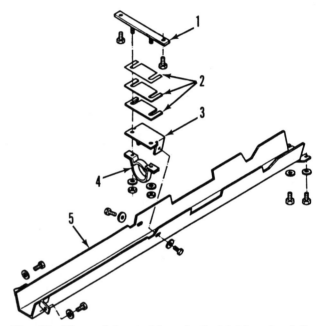

Fig. 28—View of front drive shaft shield and relative components.

1. Support bar
2. Shims
3. Shield support
4. Bearing retainer
5. Shield

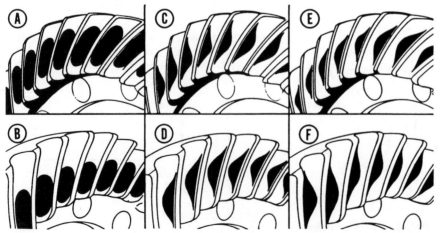

A. Proper tooth contact, coast side pattern
B. Proper tooth contact, drive side pattern
C. Pinion requires thicker shim, coast side pattern
D. Pinion requires thicker shim, drive side pattern
E. Pinion requires thinner shim, coast side pattern
F. Pinion requires thinner shim, drive side pattern

Fig. 27—Views of bevel ring gear tooth contact patterns.

All Models With Mechanical Shift

17. R&R AND OVERHAUL. To remove the transfer gearbox, first remove drive shaft shield and front drive shaft as outlined in paragraph 16. Then, remove plug (18—Fig. 29) and drain oil. Drive out roll pin and remove shift lever from gearbox Place a floor jack under gearbox, unbolt and carefully lower gearbox from tractor.

Place the gearbox in a soft jawed vise with output shaft pointing upward. Remove snap ring (6) and retainer (7). Withdraw output shaft (8) and pry out oil seal (9). Remove snap ring (10), then using an inter-

nal puller CAS-10690 and slide hammer CAS-10581 or equivalent, remove shaft (12) from gearbox. Remove bearing (11) from shaft. Remove selector collar (13), gear (14), thrust washer (26) and inner race of bearing (27). Remove pads (28), then remove washers (33), spring (31), bolt (29) and washers from both sides of shift fork (30). Drive the four roll pins (35) into shaft (34) and withdraw shaft. Remove left lever (32), detent ball (23), shift fork (30), right lever (36) and its detent ball as shaft is removed. Remove snap rings (19), plugs (20) and springs (21 and 22) from both sides of gearbox. Remove cup plugs (15) and drive roll pins (16 and 17) from both sides. Drive roll pins (35) out of shaft (34). Clean and inspect all parts and renew as necessary. If roller bearing (27) shows excessive wear or other damage, use internal puller CAS-10690 to remove.

When reassembling, use new "O" ring (37) and oil seal (9), then reassemble by reversing the disassembly procedure. With reassembly completed, adjust bolts and nuts (29 and 33) until shift pads (28) on yoke are centered in selector collar (13) as shown at (a—Fig. 30). Clearance should measure 0.0394 inch (1.0 mm) on each side when correctly adjusted. Lock the nuts together.

Reinstall transfer gearbox, using new gasket (25—Fig. 29). Tighten retaining bolts to a torque of 90 ft.-lbs. (120 N·m). Install shift lever and secure with roll pin. Reinstall front drive shaft and drive shaft shield as outlined in paragraph 16. Check transmission oil level and refill as necessary.

All Models With Electric/Hydraulic Shift

18. R&R AND OVERHAUL. To remove the transfer gearbox, first remove drive shaft shield and front drive shaft as in paragraph 16. Remove plug (17—Fig. 31) and drain oil. Disconnect electrical conduit pipe from union near solenoid cover, then remove the un-

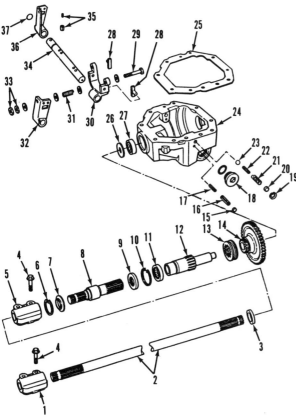

Fig. 29—Exploded view of drive shaft and mechanical shift (dog clutch) transfer case used on some models.

1. Front coupler
2. Front drive shaft
3. Ball bearing
4. Cap screws
5. Rear coupler
6. Snap ring
7. Retainer
8. Output shaft
9. Oil seal
10. Snap ring
11. Roller bearing
12. Shaft
13. Selector collar
14. Gear
15. Cup plug
16. Roll pin
17. Roll pin
18. Drain plug
19. Snap ring
20. Plug
21. Spring
22. Spring
23. Ball
24. Housing
25. Gasket
26. Thrust washer
27. Roller bearing
28. Pads
29. Bolt
30. Yoke
31. Spring
32. Lever (L.H.)
33. Nuts
34. Shaft
35. Roll pins
36. Lever (R.H.)
37. "O" ring

Fig. 30—Adjust nuts and bolts on shift mechanism until shift pads on yoke are center in selector collar as shown at "a."

ion. Remove solenoid cover cap screws (23) and remove cover (22) and gasket (21). Disconnect wire from solenoid (19). Disconnect and cap the hydraulic line at left rear of transfer gearbox. Place a floor jack under the gearbox, unbolt and carefully lower gearbox from tractor.

Place the gearbox in a soft jawed vise with output shaft pointing upward. Remove snap ring (6) and re-

tainer (7). Withdraw output shaft (8) and pry out oil seal (9). Remove snap ring (10) and insert blocks above clutch carrier (25). Use an internal puller CAS-10690 and slide hammer CAS-10581 or equivalent, and remove shaft (12). Remove disc clutch assembly (24 through 40). Remove thrust washer (15) and inner race of bearing (16). Use the internal puller to remove the bearing. Remove roller bearing (11) and seal rings (13 and 14) from shaft (12). Remove socket head screws (20) and withdraw solenoid valve (19).

To disassemble the clutch assembly, place unit in a press with Belleville springs up. Use a cutaway short piece of pipe to allow removal of snap ring (40). Then, remove spacer ring (39) and Belleville springs (38). Invert the unit and remove snap ring (24), end plate (26), clutch discs (27 and 28) and plates (29 and 30). Remove clutch piston (35) by tapping the ends of pressure pins (34) alternately. Remove seal rings (36 and 37) from piston and "O" rings (33) from pressure pins (34). Clean and inspect all parts and renew any showing excessive wear or other damage.

Use all new "O" rings, seal rings, oil seals and gaskets when reassembling. Reassemble by reversing the disassembly procedure keeping the following points in mind: Install piston seal rings (36 and 37) with larger diameter of the tapered rings toward pressure side of piston. Install new "O" rings (33) on pressure pins (34) and put grease on "O" rings and on foot plate of pins. Install pressure pins in groove of piston, lubricate seal rings and install piston with pins into gear cylinder (32). Install three Belleville springs (38) against piston and spacer ring (39) with chamfered side toward Belleville springs. Compress springs and install snap ring (40). Turn unit over and install constant thickness 0.157 inch (4.0 mm) end plate (30). Install one medium variable thickness 0.213 inch (5.4 mm) plate (29). Install shaft (12) temporarily to align clutch plate inner splines. Then, starting with an outer splined disc (28) and followed by an inner splined disc (27), alternately install all 17 discs. Install end plate (26). Press inward on clutch plates to compress Belleville springs and install snap ring (24). Continue applying approximately 2250-3375 lbs. (10000-15000 N) with press on clutch plates and measure distance from top of clutch carrier to end plate. Record the measurement. Completely release all pressure and again measure distance from top of carrier to end plate. Subtract the second measurement from the first measurement to obtain clutch disc clearance which should be within the limits of 0.031-0.071 inch (0.8-1.8 mm). If clearance is not within the limits, disassemble and change thickness of variable thickness plate (29). Plate is available in 14 thicknesses from 0.157 to 0.268 inch (4.0 to 6.8 mm). When installing seal rings on shaft (12), make certain ends of seals are hooked together. Install three new "O" rings on solenoid valve (19), lubricate the "O" rings and install solenoid valve.

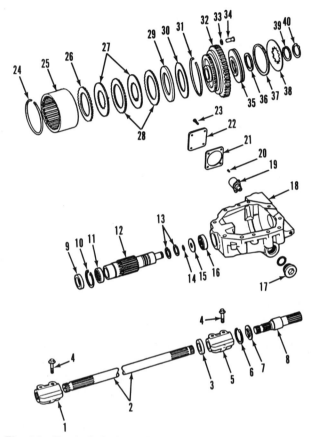

Fig. 31—Exploded view of front drive shaft and transfer case with electric/hydraulic multiple disc clutch used on some models.

1. Front coupler
2. Front drive shaft
3. Ball bearing
4. Cap screws
5. Rear coupler
6. Snap ring (internal)
7. Retainer
8. Output shaft
9. Oil seal
10. Snap ring (external)
11. Roller bearing
12. Shaft
13. Rings
14. Ring
15. Washer
16. Roller bearing
17. Drain plug
18. Housing
19. Solenoid valve
20. Socket head screw (2)
21. Gasket
22. Cover
23. Cap screw (4)
24. Snap ring
25. Carrier
26. End plate
27. Inner spline discs
28. Outer spline discs
29. Variable thickness plate
30. Constant thickness plate
31. Snap ring
32. Gear
33. "O" ring (3)
34. Pressure pins (3)
35. Piston
36. Seal ring (inner)
37. Seal ring (outer)
38. Belleville spring (3)
39. Spacer ring
40. Snap ring

When unit is reassembled, use new gasket, reinstall transfer gearbox and tighten retaining bolts to a torque of 90 ft.-lbs. (120 N·m). Connect electric wire to solenoid and hydraulic line to gearbox. Use new gasket (21) and install cover (22). Check transmission oil level and refill with Hy-Tran Plus oil.

Before installing front drive shaft, check the slip torque of the gearbox clutch as follows: Place transmission in first gear and low range and apply park brake. Using a torque wrench and special tool CAS-1876 on end of output shaft, check clutch slip torque. If unit was assembled with new clutch parts, clutch should slip at 654 ft.-lbs. (887 N·m). If used clutch parts were used, clutch should slip at 594 ft.-lbs. (806 N·m). If slip torque is considerably lower, Belleville springs may be weak and should be renewed.

Reinstall front drive shaft and drive shaft shield as outlined in paragraph 16.

TIE RODS AND TOE-IN

All Models With Front Drive Axle

19. Toe-in must be 0-3/16 inch (0-5 mm) and is adjusted by turning the threaded studs between steering cylinder rod and tie rod ends. Both sides must be adjusted equally.

POWER STEERING SYSTEM

NOTE: The maintenance of absolute cleanliness of all parts is of utmost importance in the operation and servicing of the hydrostatic steering system. Of equal importance is the avoidance of nicks or burrs on any of the working parts.

STEERING CIRCUITS

All Models

20. OPERATION. Power steering is standard equipment on all models. The internal hydraulic pump (2—Fig. 32) pulls oil from the reservoir through a full flow suction filter installed in the outside of the multiple control valve (13). The filtered oil is pressurized by the pump and is supplied to the priority flow divider valve in the multiple control valve. The flow divider valve directs 3 gpm (11.35 L/min.) of oil to the steering system at all engine speeds. The fluid enters the hand pump and flows through the 2150 psi (14824 kPa) relief valve. With steering wheel in neutral position, oil flows out of return port and back to the multiple control valve. If steering wheel is turned in either direction, fluid in the hand pump will be directed through the metering valve to the desired steering cylinder port. Return oil from the cylinder flows back through the hand pump and on to the multiple control valve.

Some of the steering return oil is used to operate the power shift transmission and the front drive axle clutch valve, if so equipped. The remainder of the return oil is directed to the operating spool for the pto clutch and to the transmission oil cooler. Oil from the cooler is used to supply the brake system.

LUBRICATION AND BLEEDING

All Models

21. The tractor rear frame serves as a common reservoir for all hydraulic and lubrication operations.

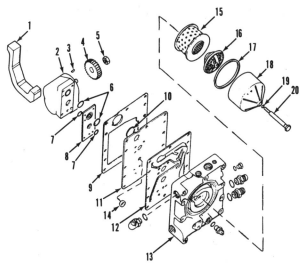

Fig. 32—Exploded view showing hydraulic pump, filter, multiple control valve and relative components.

1. Foam seal
2. Hydraulic pump
3. Woodruff key
4. Gear
5. Nut
6. "O" rings
7. "O" rings
8. Spacer plate
9. Inner gasket
10. "O" ring
11. Plate
12. Outer gasket
13. Multiple control valve
14. Seal ring
15. Filter element
16. Bypass valve
17. Gasket
18. Filter case
19. "O" ring
20. Center bolt

The filter element (15—Fig. 32) should be renewed after 20 hours, 200 hours and then every 800 hours of operation. Hydraulic fluid should be drained and new fluid installed every 800 hours of operation or once a year, whichever occurs first. Only Hy-Tran Plus fluid should be used and level should be maintained at full mark on dipstick. Refill capacity is 36 U.S. quarts (34 L) on tractors without front drive axle and 38.6 U.S. quarts (36.5 L) on tractors with front drive axle.

Whenever power steering lines have been disconnected or the fluid changed, start engine and cycle power steering system from stop to stop several times to bleed air from system. Then, check and if necessary, add transmission fluid.

HYDROSTATIC STEERING TESTS

All Models

IMPORTANT: Before any hydraulic tests are made, transmission oil in the tractor must be warmed to a temperature of at least 100° F (38° C).

22. PRIORITY FLOW DIVIDER. To check the flow divider valve (33—Fig. 33), first remove the pilot relief valve (29) and install a test pilot relief valve from test kit No. 399858R92. Disconnect power steering supply pipe (directly below hydraulic filter). Cap the pipe and connect the inlet hose of a flowmeter to the steering supply union. Place the flowmeter outlet hose in the transmission oil fill hole and secure with wire. Move the position and draft control levers fully forward to prevent hitch movement during test. Fully open load valve on flowmeter. Start and operate engine at low idle speed. Turn load valve until a reading of 1000 psi (6895 kPa) is on pressure gage and record the flow. Increase the load to 1600 psi (11032 kPa) and record the flow. Fully close the load valve and record the reading on pressure gage. Then, fully open the load valve. Increase the engine speed to 2180 rpm on Models 385 and 485, 2300 rpm on Model 585 and 2400 rpm on Models 685 and 885. Repeat the steering flow checks and maximum pressure test at the correct rpm for each model. All of the flow readings should be between 2.6 and 3.1 gpm (9.9 and 11.8 L/min.). Maximum pressure with the load valve fully closed must be 2500 to 2600 psi (17238 to 17927 kPa).

If flow or pressure readings are not correct, check for the following conditions:

1. Excessive wear on flow divider spool.
2. Faulty flow divider spring.
3. Faulty pilot relief valve seal rings.
4. Clogged filter or faulty pump.

Correct the problem and retest. Remove test pilot relief valve and using new seal rings, install original pilot relief valve. Remove flowmeter and connect steering supply line.

23. STEERING HAND PUMP. To test the steering hand pump, disconnect steering hose from left end of steering cylinder. Cap cylinder fitting. Connect a hose and 4000 psi (28000 kPa) test gage to the steering hose. Remove center cap from steering wheel.

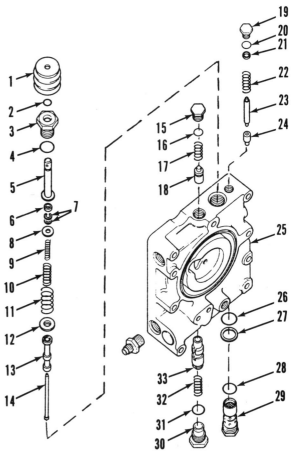

Fig. 33—Exploded view of the multiple control valve assembly.

1. Boot
2. "O" ring
3. Guide
4. "O" ring
5. Stem
6. Spacer
7. Shims
8. Washer
9. Spring (heavy)
10. Spring (light)
11. Valve return spring
12. Retainer
13. Valve spool
14. Guide pin
15. Plug
16. "O" ring
17. Spring
18. Oil collar bypass valve
19. Plug
20. "O" ring
21. Shim
22. Spring
23. Pin
24. Pto pressure regulator valve
25. Multiple control valve housing
26. "O" ring
27. Back-up ring
28. "O" ring
29. Pilot relief valve
30. Plug
31. "O" ring
32. Spring
33. Flow divider spool

With engine stopped, use a torque wrench and apply 52 ft.-lbs. (70 N·m) of torque to the steering wheel nut in a counterclockwise direction. Check the reading on the test gage. Pressure reading should be not less than 55 psi (380 kPa). If pressure reading is not correct, check for the following conditions:

1. Worn steering hand pump or damaged seals.
2. Worn or stuck check ball in hand pump.
3. Stuck check valve in pressure inlet port of hand pump.

Then, with engine operating at 1000 rpm, turn steering wheel counterclockwise and check pressure gage reading. Pressure should read 2150 psi (14825 kPa). If pressure reading is not correct, check for following conditions:

1. Relief valve stuck in hand pump.
2. Incorrect relief valve setting. (Paragraph 29).

Correct the problem and retest. Remove pressure gage and connect steering hose to cylinder.

24. STEERING CYLINDER SEALS. To check steering cylinder seals, operate engine at low idle speed. Turn steering wheel from lock to lock at least four times to remove all air from the system. Turn wheels fully to the left and stop the engine. Disconnect steering hose from left end of steering cylinder. Connect a high pressure cap on steering hose. Allow any oil to flow from the open union on steering cylinder. Start engine and operate at low idle speed. Check for oil leaking from open union and at the right steering shaft seals in end cap. A small amount of oil (up to 5 drops in 30 seconds) is permissible for leakage past the piston seals. There should be no leakage at shaft seals in end cap. If cylinder tests OK, repeat the test on right end of cylinder.

25. HYDRAULIC PUMP. To check flow and pressure of the hydraulic pump, first remove the plug (30—Fig. 33) for the flow divider. Remove spring (32) and spool (33), then reinstall plug. Remove the pilot relief valve (29) and install a test pilot relief valve from test kit No. 399858R92. Disconnect the power steering supply pipe (directly below hydraulic filter). Cap the pipe and union on multiple control valve with high pressure caps. Disconnect hitch supply line (bottom rear line on multiple control valve) and cap the line. Connect the inlet hose from the flowmeter to the hitch supply union on the multiple control valve. Place outlet hose from flowmeter into transmission oil fill hole and wire in place. Fully open load valve on flowmeter. Start engine and operate at a speed of 2180 rpm for Models 385 and 485, 2300 rpm for Model 585 or 2400 rpm for Models 685 and 885. Record the flow. Close load control valve to obtain

a reading of 2000 psi (13790 kPa) on pressure gage, then record the flow. Fully open the load valve and stop engine. The second flow reading must not be less than 90 percent of the first flow reading. If second reading is too low, check for the following conditions:

1. Low oil level.
2. Clogged filter.
3. Faulty pump.

Remove test pilot relief valve and using new seal rings, install original pilot relief valve. Remove flow divider plug (30) and install spool (33) and spring (32). Reinstall and tighten plug. Remove caps from steering supply line and union, then connect steering line. Disconnect and remove flowmeter and connect hitch supply line. Start tractor and bleed air from steering system.

HYDRAULIC PUMP

All Models

26. Tractors may be equipped with either Sunstrand or Plessy hydraulic gear type pumps. The maximum pump output is controlled by engine rpm and the number of teeth on the pump drive gear. Refer to chart in Fig. 34 for the model, engine test speed, available drive gear teeth and maximum total pump output. The priority flow divider sends approximately 3 gpm to the steering system and the balance is used for the draft and position controls on the hitch lift system and for the auxiliary control valves.

27. R&R AND OVERHAUL. To remove the hydraulic pump, first drain all compartments of Hy-Tran Plus fluid.

NOTE: Although the pump can be removed with left rear wheel installed, most mechanics prefer to remove left rear wheel to provide more room to remove the multiple control valve and pump assembly.

Thoroughly clean the area around the multiple control valve, then remove the hydraulic filter assembly

Tractor Model	Engine Speed RPM	Drive Gear No. of Teeth	Total Pump Output	
			U.S. Gals	Litres
385	2180	21	15.6	59.1
		20	16.4	62.2
485	2180	21	15.6	59.1
		20	16.4	62.2
585	2300	23	15.1	62.5
		20	17.3	65.6
685	2400	23	15.7	59.5
		20	18.1	68.6
885	2400	23	15.7	59.5
		20	18.1	68.6

Fig. 34—Chart showing engine test speed for each model, number of teeth available on pump drive gear and maximum output of pump.

(15 through 20—Fig. 32). Disconnect steering lines, oil cooler line, hitch supply line, auxiliary valves return line and pto line from multiple control valve (13). Immediately cap or plug all openings. On models equipped with XL cab, remove left console. On all models, disconnect and remove park brake warning switch wires and linkage. Disconnect and remove pto operating lever. Remove two cap screws from opposite sides of multiple control valve. Install two 3/8 × 5 inch guide studs in the holes to aid in removal and installation of the assembly. Remove the balance of the retaining cap screws, then slide multiple control valve outward from rear main frame. Lay the assembly on a clean bench. Remove the four cap screws and copper washers, then lift off hydraulic pump assembly (2) and spacer plate (8), if so equipped.

NOTE: Spacer plate (8) is used only on tractors equipped with a 23 tooth pump drive gear.

Disassembly of the pump is obvious after examination of the unit. Remove nut (5), drive gear (4) and Woodruff key (3). Remove end plate cap screws and separate pump components. Clean and inspect all parts. If any parts show excessive wear or other damage, renew pump assembly. A seal kit is available for resealing the pump. When reassembling pump, tighten end plate cap screws to a torque of 20-23 ft.-lbs. (27-31 N·m). Install Woodruff key (3), place drive gear (4) on shaft and install and tighten nut (5) to a torque of 45-50 ft.-lbs. (61-68 N·m). Hold drive gear in a vise while tightening nut. Check condition of foam seal (1). If damaged or loose, remove and discard seal. Clean seal area with a rag and adhesive solvent, then dry the housing. Apply a thin coat of oil resistant adhesive No. 69351C1 to the housing and to a new foam seal. Let adhesive dry one to two minutes, then install foam seal on pump housing. Allow adhesive to dry at least another 30 minutes before installing pump assembly.

When reassembling, use all new "O" rings, seal ring and gaskets. Install pump and spacer (8), if so equipped, to multiple control valve. Use new copper washers on the four pump retaining cap screws. Install and tighten cap screws to a torque of 18-24 ft.-lbs. (24-28 N·m).

Lubricate foam seal (1) with Hy-Tran Plus fluid and reinstall multiple control valve and pump assembly by reversing the removal procedure. Tighten multiple control valve retaining cap screws to a torque of 33-37 ft.-lbs. (45-50 N·m). Reinstall hand park brake linkage and pto linkage. Then, using a new filter element (15), "O" rings and gasket, install hydraulic filter assembly. Tighten filter center bolt to a torque of 12-16 ft.-lbs. (16-21 N·m). Connect all disconnected hydraulic lines. Fill transmission to correct level on dipstick with Hy-Tran Plus oil. Refill capacity is 36 U.S. quarts (34.0 L) on tractors without front drive axle and 38.6 U.S. quarts (36.5 L) on tractors with front drive axle.

PRIORITY FLOW DIVIDER

All Models

28. R&R AND OVERHAUL. To remove the flow divider spool (33—Fig. 33) and spring (32), remove plug (30) with "O" ring (31). Clean and inspect parts. Spring (32) should have a free length of 3.253 inches (82.63 mm) and should test 29.7 lbs. (132 N) when compressed to a length of 2.017 inches (51.23 mm). Check spool (33) and spool bore in multiple control valve housing (25) for excessive wear, scoring or other damage. Spool (33) and housing (25) are a matched set and are renewed only as a new multiple control valve assembly.

Lubricate spool with Hy-Tran Plus fluid, then install spool, spring and plug with new "O" ring. Tighten plug securely.

STEERING HAND PUMP

All Models

29. R&R AND OVERHAUL. To remove the steering hand pump, proceed as follows: On models without cab, remove battery access panel and disconnect battery cables. Unbolt and remove upright muffler. Remove radiator cap, tractor hood and side panels. Remove battery and insulation pad from battery tray. Using a suitable puller, remove steering wheel. Remove nut from ignition switch and push switch to inside. Unbolt and remove instrument panel side panels. Unbolt instrument panel and raise about 2 inches (50 mm) on right side and support in this position.

On models with XL cab, remove battery cover, disconnect battery cables and remove battery. Using a suitable puller, remove steering wheel. Push in and turn lockpins, then lift off the instrument panel front housing. Loosen locknut and remove engine stop control knob. Remove locknut and grommet from instrument panel. Disconnect tachometer cable. Lift instrument panel upward to clear steering shaft, then move the panel to the right and support assembly in this position.

On all models, identify and disconnect the four steering hoses from hand pump. Plug or cap all openings. Remove the two hand pump retaining cap screws and lower the unit. Remove the two spacers, then remove hand pump assembly.

To disassemble the steering hand pump, clean exterior of assembly, then unbolt and remove steering shaft and column. Scribe match marks across end cover (32—Fig. 35), stator (30), distributor plate (27) and housing (9). Clamp valve housing in a soft jawed vise in inverted position and remove end cover cap screws. Remove end cover (32), stator (30), spacer (25), rotor (28), distributor plate (27) and drive shaft (24). Using a screwdriver and a magnet, unscrew and remove threaded bushing (14) and ball (13). Remove unit from vise and shake suction pins (11) and balls (10) from housing. DO NOT use a magnet to remove pins and balls. Push valve spool assembly from housing. Remove thrust washers (16 and 18), needle bearing (17) and retaining ring (19) from end of spool. Remove cross pin (21) from spool, then slide spool (20) from sleeve (22). Remove inner and outer plate

springs (23). Remove seal ring (15), "O" ring (12) and dust seal (8) from housing. Using an Allen wrench, remove plug (1) with "O" ring (2). Note the position of the adjusting plug (3), then unscrew while counting the number of turns needed to remove the adjusting plug. Record the number of turns to aid in reassembly. Remove spring (4) and relief valve (5). DO NOT remove seat (6) or check valve (7) as they are not available as replaceable parts.

Clean and inspect all parts for excessive wear or other damage. If housing (9), seat (6), check valve (7), spool (20), sleeve (22), rotor (28) or stator (30) are not suitable for further service, renew complete hand pump assembly as these parts are not serviced separately. All "O" rings, seals and plate springs are available in a seal kit. Check rotor (28) and stator (30) for wear as follows: Refer to Fig. 36 and measure thickness of rotor at (A) and stator at (B). If (A) is 0.002 inch (0.051 mm) less than (B), renew hand pump. Refer to Fig. 37 and place rotor in stator as shown. Using a feeler gage, measure gap as shown. If gap is 0.005 inch (0.127 mm) or more, renew hand pump assembly.

Reassemble by reversing the disassembly procedure keeping the following points in mind. Lubricate all internal parts with Hy-Tran Plus fluid and coat all "O" rings with petroleum jelly during reassembly. When installing plate springs (23—Fig. 35), first install the two flat springs. Then, install the two curved springs (arches back to back) in between the two flat springs. When installing rotor (28) on shaft (24), make certain that cross pin slot in shaft aligns with center of a valley in rotor. Align match marks and tighten end cover cap screws (35) to a torque of 22-26 ft.-lbs. (30-35 N·m).

Reinstall steering column and upper shaft on pump and tighten cap screws to a torque of 33-37 ft.-lbs. (45-50 N·m). Install hand pump and two spacers in place, then install the two retaining cap screws and torque to 33-37 ft.-lbs. (45-50 N·m). With pump installation completed, install all removed parts (bat-

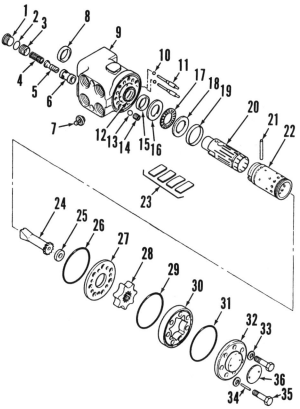

Fig. 35—Exploded view of the Danfoss steering hand pump.

1. Plug	19. Retaining ring
2. "O" ring	20. Spool
3. Adjusting plug	21. Cross pin
4. Spring	22. Sleeve
5. Relief valve	23. Plate springs
6. Seat	24. Drive shaft
7. Check valve	25. Spacer
8. Dust seal	26. "O" ring
9. Housing	27. Distributor plate
10. Ball (2)	28. Rotor
11. Suction valve pin (2)	29. "O" ring
12. "O" ring	30. Stator
13. Check ball	31. "O" ring
14. Threaded bushing	32. End cover
15. Seal ring	33. Washer (7)
16. Thrust washer	34. Roll pin
17. Needle bearing	35. Special cap screw
18. Thrust washer	36. Name plate

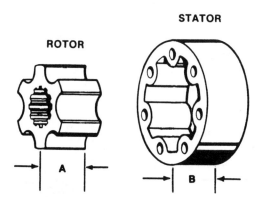

Fig. 36—When checking rotor and stator, measure distance "A" on rotor and distance "B" on stator. Refer to text.

tery, hood, panels and instrument panel) except access panel to adjust relief valve (5) pressure. Refer to paragraph 23 and adjust relief valve pressure as follows: Remove plug (1—Fig. 35) and with engine operating at 1000 rpm, check relief pressure on gage. Pressure should be 2150 psi (14825 kPa). Turn adjusting plug (3) inward to increase pressure or outward to decrease pressure. Install and tighten plug (1) and complete balance of installation.

STEERING CYLINDER

Various types of steering cylinders are used depending on front axle type. Refer to the appropriate following paragraphs for overhaul procedures.

Models With Adjustable Front Axle

30. R&R AND OVERHAUL. To remove the steering cylinder, first disconnect steering hose at each end of cylinder. Plug or cap all openings. Remove the clevis pin at right end of cylinder shaft. Straighten lockstrip and unbolt lower pivot arm. Remove lower arm, then remove clevis end from anchor on axle main member. Lower cylinder assembly from upper pivot arm.

To disassemble the cylinder, first clean exterior of cylinder and clamp lower pivot flange in a soft jawed vise. Remove elbow fitting (11—Fig. 38) from left end of cylinder rod and clevis (15) from right end. Lift end of retaining ring (13) out of slot, then using a pin type spanner, rotate cylinder head (6) and work retaining ring out of its groove. Cylinder rod and piston assembly (3) can now be removed from cylinder tube (12). Remove remaining cylinder head in the same manner. All seals, "O" rings and back-up rings are now available for inspection and/or renewal. Clean all parts in a suitable solvent and inspect. Check cylinder tube (12) for scoring, grooving and out-of-roundness. Light scoring can be polished out by using a fine emery cloth and oil, providing a rotary motion is used during polishing operation. A cylinder

tube that is heavily scored or grooved, or is out-of-round, should be renewed, Check piston rod and piston (3) for scoring, grooving and straightness. Polish out very light scoring with fine emery cloth and oil, using a rotary motion. Renew rod and piston assembly if heavily scored or grooved, or if piston rod is bent. Inspect bores of cylinder heads (6) and renew same if excessively worn or out-of-round. All "O" rings, seals, back-up rings and retainer rings are available in a seal kit. Lubricate all "O" rings, back-up rings and seals during assembly.

Reassemble steering cylinder as follows: Install a new "O" ring (2) in groove of piston, then follow with new seal ring (1). Install "O" ring (7), back-up ring (8) and wiper seal (9) in each cylinder head (6) and back-up ring (5) and "O" ring (4) over each cylinder head. Using a ring compressor or suitable hose clamp around piston seal ring (1), install piston and rod assembly. Install cylinder head assemblies over each end of piston rod and into cylinder tube. Align hole in groove of cylinder heads so they will accept nib on end of retaining rings (13). Rotate cylinder heads until retaining rings are pulled completely into cylinder tube. Using new "O" rings (10 and 14), install elbow (11) in left end of cylinder rod and clevis (15) on right end. Tighten clevis to a torque of 170-180 ft.-lbs. (230-244 N·m).

Reinstall cylinder on tractor by reversing removal procedure. Tighten lower arm retaining cap screws

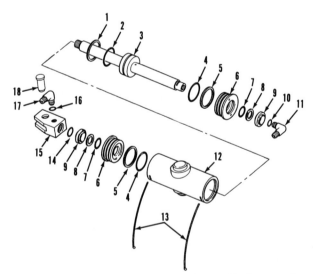

Fig. 38—Exploded view of steering cylinder used on models with adjustable front axle.

1. Seal ring	
2. "O" ring	
3. Rod & piston assy.	10. "O" ring
4. "O" ring	11. Elbow fitting
5. Back-up ring	12. Cylinder tube
6. Cylinder head	13. Retaining rings
7. "O" ring	14. "O" ring
8. Back-up ring	15. Clevis
9. Wiper seal	16. "O" ring
	17. Elbow fitting
	18. Pin

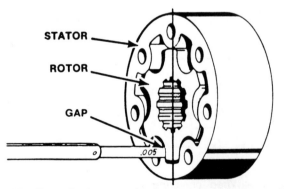

Fig. 37—Use a feeler gage to measure gap between rotor and stator.

to a torque of 80-90 ft.-lbs. (110-124 N·m). Bend corners of lock plate up against flat on cap screw heads. Bleed air from power steering as outlined in paragraph 21.

Models With Nonadjustable (Cast) Front Axle

31. R&R AND OVERHAUL. To remove the steering cylinder, first disconnect steering hose at each end of cylinder. Disconnect steering cylinder ball joint end from left steering arm. Remove pivot pin at right end of cylinder. Then, lift cylinder from tractor. Drain oil from cylinder, then thoroughly clean exterior of cylinder.

To disassemble steering cylinder, clamp the flat end of steering tube (1—Fig. 39) in a soft jawed vise so cylinder is in vertical position. Remove small slot head lock screw (9) and lockwasher from cylinder. Use a filter wrench CAS-04-535 or equivalent, rotate cylinder counterclockwise until the straight end of retaining ring (5) can be seen through slot in cylinder. Pry end of retaining ring out through slot and

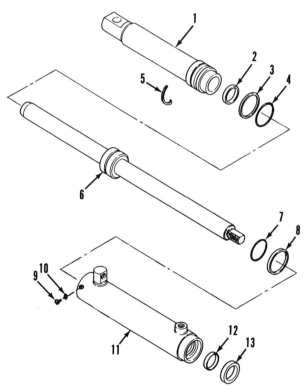

Fig. 39—Exploded view of steering cylinder used on models equipped with nonadjustable (cast) front axle.

1. Steering tube	7. "O" ring
2. Seal ring	8. Seal ring
3. Back-up ring	9. Lock screw
4. "O" ring	10. Lockwasher
5. Retaining ring	11. Cylinder
6. Rod & piston	12. Seal ring
assy.	13. Wiper seal

continue to turn cylinder counterclockwise until ring is removed. Pull steering rod and cylinder assembly from the tube (1). Remove seal ring (2), back-up ring (8) and "O" ring (4) from end of steering tube. Place steering rod ball joint in the vise, remove clamping bolt, then remove ball joint. Lightly clamp cylinder in the vise and pull rod and piston (6) from cylinder. Remove wiper seal (13) and seal ring (12) from cylinder and seal ring (8) and "O" ring (7) from piston.

Clean and inspect all parts. Inspect cylinder (11) for scoring or out-of-roundness. Light scoring can be polished out using a fine emery cloth and oil, providing a rotary motion is used during polishing operation. Renew cylinder if heavily scored or out-of-round. Check piston rod and piston (6) for scoring and straightness. Polish out light scoring with fine emery cloth and oil, using a rotary motion. Renew rod and piston assembly if heavily scored or if rod is bent. All "O" rings, back-up rings, seal rings and wiper seal are available in a seal kit. Lubricate all internal parts with Hy-Tran Plus fluid during reassembly.

Reassemble steering cylinder by reversing the disassembly, keeping the following points in mind: Heat Teflon seal rings and back-up rings in clean Hy-Tran Plus oil to a temperature of 75° F (24° C) to make them more flexible. Tighten clamping bolt on ball joint to a torque of 60-65 ft.-lbs. (81-85 N·m).

For aid in installation, set distance from center of pivot pin hole to center of ball joint to 35-1/2 inches (905 mm). With front wheels in straight ahead position, reinstall cylinder assembly. Tighten ball joint nut to a torque of 50-100 ft.-lbs. (68-136 N·m) and secure with cotter pin. When installation of cylinder is complete, bleed air from power steering as outlined in paragraph 21.

Models With Front Drive Axle

32. OVERHAUL. The power steering cylinder is contained in the back side of front drive axle housing. See Fig. 40. Cylinder can be disassembled without removing axle assembly.

Disconnect steering lines at each end of cylinder. Cap or plug all openings immediately. Clean exterior of cylinder and surrounding area. Disconnect tie rod ends from swivel housing arms. Unbolt and remove steering stop collar at each end of cylinder rod. Remove tie rods by unscrewing threaded studs (3) from ends of cylinder rod. Unbolt and remove right side guide (22—Fig. 41) and shim spacer (20). Withdraw piston and rod assembly, then remove cylinder sleeve (6) and left side guide (4). Remove scraper (2) and groove rings (3) from both end guides. Remove the two Teflon rings (8) and piston seal ring (9) from piston. Remove snap ring (10), collar ring (11) and split ring (12), raise, then lower piston and remove spacer (13) and "O" ring (14). Remove piston (15), spacer (16), split ring (17), collar ring (18) and snap ring (19).

Clean and inspect all parts for excessive wear or other damage. Inspect cylinder sleeve (6), piston (15), rod (23) and rod guides (4 and 22) for cracks, scoring or other damage. Light scoring can be polished out using fine emery cloth and oil, providing a rotary motion is used during the polishing procedure. Use all new "O" rings, seals and seal rings and lubricate with clean Hy-Tran Plus fluid during reassembly.

Reassemble cylinder by reversing disassembly procedure and adjust cylinder sleeve to zero end play as follows: With lips of grooved ring (3) and scraper ring (2) facing inward and "O" ring (1) in external groove, install left rod guide (4) with notch facing hydraulic port. With "O" rings (5 and 7) installed in grooves of cylinder sleeve, install sleeve until fully seated on left rod guide. Measure distance between outer and inner faces of right side rod guide as shown in Fig. 42 and record the distance. Then, measure distance between axle housing and end of cylinder sleeve as shown in Fig. 43 and record the distance. Subtract the second distance from the first distance to determine correct thickness of shim spacer (20—Fig. 41) to be used. Shim spacers are available in thicknesses of 0.035 to 0.071 inch (0.9 to 1.8 mm) in increments of 0.004 inch (0.1 mm). Install rod and piston assembly in cylinder sleeve. With lips of grooved ring (3) and scraper ring (2) facing inward in right side rod guide, use vaseline to stick the determined shim spacer to inside of rod guide. Install right side rod guide and tighten rod guide cap screws to a torque of 52 ft.-lbs. (70 N·m).

Apply Loctite 242 to the first three threads and install threaded studs (3—Fig. 40) in cylinder rod ends.

Tighten securely. Install tie rod ball joints in swivel housing arms and tighten nuts to a torque of 74-81 ft.-lbs. (100-110 N·m). Remove plugs from steering cylinder, fill cylinder with clean Hy-Tran Plus fluid, then connect steering lines. Install steering stops on each end of steering cylinder rod. Bleed air from system as outlined in paragraph 21. Check and adjust toe-in, if necessary, as outlined in paragraph 19.

OIL COOLER

All Models

33. Service of the oil cooler involves only removal and reinstallation, or renewal of faulty units. However, since it is hinged on lower side, the oil cooler will swing down for cleaning. Removal of oil cooler is obvious after examination of the unit. Inlet and outlet hoses must be identified as they are removed from cooler pipes so oil circuits will be kept in the proper sequence.

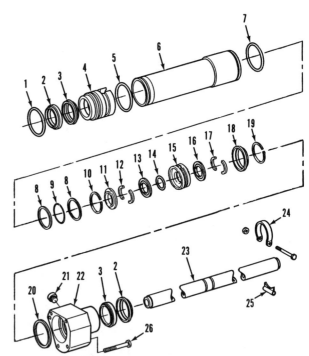

Fig. 41—Exploded view of steering cylinder used on models equipped with front drive axle.

1. "O" ring	14. "O" ring
2. Scraper ring	15. Piston
3. Grooved ring	16. Spacer
4. Rod guide (left side)	17. Split ring
	18. Collar ring
5. "O" ring	19. Snap ring
6. Cylinder sleeve	20. Shim spacer
7. "O" ring	21. Plug
8. Teflon rings	22. Rod guide (right side)
9. "O" ring	23. Rod
10. Snap ring	24. Steering stop
11. Collar ring	25. Spacer plate
12. Split ring	26. Cap screw
13. Spacer	

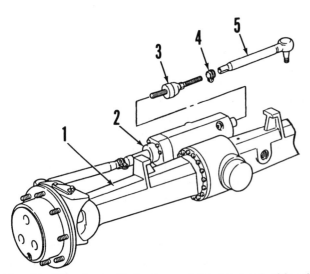

Fig. 40—Steering cylinder is contained in back side of front drive axle housing.

1. Front drive axle
2. Steering cylinder
3. Threaded stud
4. Clamp
5. Tie rod (left side)

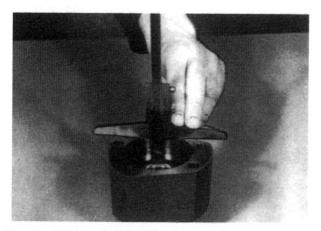

Fig. 42—Using a depth gage, measure distance between outer and inner faces of right side rod guide. Refer to text.

Fig. 43—Using a depth gage, measure distance between axle housing and end of cylinder sleeve. Refer to text.

ENGINE AND COMPONENTS

Model 385 tractors are equipped with a D-155 three cylinder diesel engine having a bore of 3.875 inches (98.4 mm) and stroke of 4.37 inches (111.1 mm) and displacement of 155 cubic inches (2.54 L).

Model 485 tractors are equipped with a D-179 three cylinder diesel engine having a bore of 3.875 inches (98.4 mm) and stroke of 5.06 inches (128.5 mm) and displacement of 179 cubic inches (2.93 L).

Model 585 tractors are equipped with a D-206 four cylinder diesel engine having a bore of 3.875 inches (98.4 mm) and stroke of 4.37 inches (111.1 mm) and displacement of 206 cubic inches (3.38 L).

Model 685 tractors are equipped with a D-239 four cylinder diesel engine having a bore of 3.875 inches (98.4 mm) and stroke of 5.06 inches (128.5 mm) and displacement of 239 cubic inches 3.92 L).

Model 885 tractors are equipped with a D-268 four cylinder diesel engine having a bore of 3.937 inches (100.0 mm) and stroke of 5.51 inches (140 mm) and displacement of 268 cubic inches (4.39 L).

R&R ENGINE ASSEMBLY

All Models

34. To remove the engine assembly, first disconnect battery cables and remove battery. Unbolt and remove hood and side panels. Drain cooling system. Identify and disconnect oil cooler lines, then remove oil cooler. Cap or plug all openings. On models so equipped, remove air conditioning condenser and tie condenser and hoses back out of the way. Unbolt and remove lower radiator seal strip. Remove wiring har-

ness from fan shroud and lay rearward. Remove radiator side plate bolts, then unbolt and remove upper radiator air baffle. Remove upper and lower radiator hoses. Unbolt and remove radiator.

If so equipped, unbolt and remove front weights and weight bracket. Remove front axle assembly as outlined in paragraph 4, 5 or 8 and front support as in paragraph 7 or 9. Identify and disconnect heater hoses from cab, if so equipped. Remove fan, fan belt and air conditioner compressor drive belt. Unbolt and remove air cleaner assembly. Identify and disconnect high and low pressure switches at compressor. Unbolt compressor mounting bracket and secure compressor back out of the way to tractor. Disconnect tachometer cable. Disconnect throttle and fuel shut-off cables at injection pump. Disconnect and cap fuel supply and fuel return pipes. Disconnect starter motor wiring. Unbolt brake pipe support bracket. Identify, disconnect and remove steering pipes at right side of engine. Cap all openings.

Support tractor under transmission speed range gearbox. Install a lift eye to rear of engine and attach a sling and hoist to lift eyes. Remove clutch housing bolts and carefully remove engine.

Reinstall engine by reversing removal procedure. Use aligning dowels so engine will be guided straight together. Tighten clutch housing bolts to a torque of 80 ft.-lbs. (108 N·m). Fill cooling system to correct level. Bleed air from fuel system. Bleed air from steering system.

CYLINDER HEAD

All Models

35. **REMOVE AND REINSTALL.** To remove the cylinder head, disconnect battery cables and drain

coolant. Remove upright muffler and hood. Disconnect upper radiator hose from engine. Unbolt and remove air cleaner assembly. Remove breather pipe from rocker arm cover and remove exhaust elbow. Disconnect, unbolt and remove coolant filter/conditioner, if so equipped. Unbolt and remove exhaust manifold and coolant manifold. Disconnect and remove injector lines from fuel injectors and injection pump. Cap or plug all openings to prevent dirt or other foreign material from entering system. Disconnect return fuel line and ether starting line. Unbolt and remove intake manifold. Remove retaining bolts from rocker arm cover and lift off cover. Unbolt and remove rocker arms and shaft assembly, then remove push rods. Remove head bolts and lift off cylinder head.

CAUTION: The injector nozzle assemblies protrude slightly through the combustion side of cylinder head. Be extremely careful when removing or reinstalling cylinder head with injectors installed, so as not to damage injector ends. DO NOT place cylinder head on bench with combustion side down.

Clean all gasket sealing surfaces of cylinder head and check for cracks or other damage. Using a straightedge and feeler gage, check cylinder head for warpage. If warpage is more than 0.0047 inch (0.12 mm) resurface cylinder head. A maximum of 0.30 inch (0.77 mm) can be removed from cylinder head. If cylinder head height is less than 3.8795 inches (98.54 mm) after head is machined, renew head.

When reinstalling cylinder head, use new head gasket and make sure gasket sealing surfaces of cylinder block and head are clean. Use guide studs when installing head. Tighten cylinder head retaining cap screws in three steps using the sequence shown in Fig. 44 or 45. Tighten the cap screws to a torque of 30 ft.-lbs. (41 N·m) in first step, 60 ft.-lbs. (81 N·m) in second step and 110 ft.-lbs. (149 N·m) in the third step. Adjust valve tappet gap as outlined in paragraph 37.

NOZZLE SLEEVES

All Models

36. REMOVE AND REINSTALL. The cylinder head is fitted with brass injector nozzle sleeves which pass through the coolant passages. The nozzle sleeves are available as service items.

NOTE: Injector nozzle sleeves can be renewed without removing the cylinder head. However, if cylinder head is not removed, be sure to drain coolant below level of cylinder head before removing the sleeves.

To remove nozzle sleeves, disconnect injector lines and remove injectors. Use special bolt PLT-509-11 and

turn it into sleeve. Attach a slide hammer puller and remove nozzle sleeve. See Fig. 46. Use caution during this operation not to damage the sealing areas in cylinder head.

NOTE: Under no circumstances should screwdrivers, chisels or other such tools be used in an attempt to remove injector nozzle sleeves.

When installing nozzle sleeves, be sure the sealing areas are completely clean and free of scratches. Apply a light coat of Grade B Loctite on sealing surfaces of nozzle sleeves (Fig. 47). Then, using installing tool 3055344R1 drive nozzle sleeves into their bores until they bottom as shown in Fig. 48. Injector nozzle sleeves have a slight interference fit in their bores. When installing sleeves, make certain sleeve is driven straight with its bore.

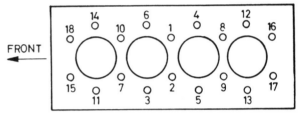

Fig. 44—Tighten cylinder head cap screws on all three cylinder models in the sequence shown.

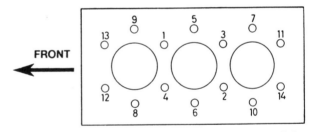

Fig. 45—Tighten cylinder head cap screws on all four cylinder models in the sequence shown.

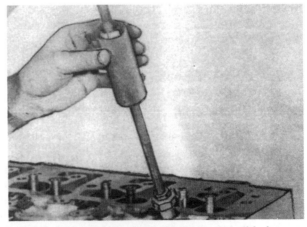

Fig. 46—Use special bolt PLT-509-11 and slide hammer to remove injector nozzle sleeve.

VALVES AND SEATS

All Models

37. Intake and exhaust valves are not interchangeable. Intake valves seat directly in the cylinder head and exhaust valves seat on renewable seat inserts. Inserts are available in oversizes of 0.006 inch (0.15 mm) and 0.016 inch (0.40 mm). Valve seat angle for both intake and exhaust valves is 44 degrees. Valve face angle for both intake and exhaust valves is 45 degrees. This provides a 1 degree interference angle between valves and seats. Valve stem seals are used on all valves and valve rotators are used on all valves.

When removing valve seat inserts, use the proper seat puller or use a pre-cup puller and expanding screw. DO NOT attempt to drive a chisel or similar tool under insert as the counterbore will be damaged. Freeze new valve seat inserts to a temperature of -76° F (-60° C) before installing.

Check the valves and seats against the following specifications:

Intake

Valve face angle	45°
Valve seat angle	44°
Stem diameter	0.3920-0.3924 in.
	(9.957-9.967 mm)
Stem to guide diametral clearance	0.0016-0.0025 in.
	(0.041-0.064 mm)
Seat width	0.060-0.070 in.
	(1.52-1.78 mm)
Valve tappet gap (warm)	0.012 in.
	(0.304 mm)
Valve recession from face of cylinder head	0.040-0.050 in.
	(1.02-1.27 mm)

Exhaust

Valve face angle	45°
Valve seat angle	44°
Stem diameter	0.3912-0.3915 in.
	(9.936-9.945 mm)
Stem to guide diametral clearance	0.0025-0.0033 in.
	(0.064-0.084 mm)
Seat width	0.060-0.070 in.
	(1.52-1.78 mm)
Valve tappet gap (warm)	0.012 in.
	(0.304 mm)
Valve recession from face of cylinder head	0.047-0.060 in.
	(1.19-1.52 mm)

CAUTION: Due to close clearance between valves and pistons, severe damage can result from inserting feeler gage between valve stem and valve lever (rocker arm) with engine running. DO NOT attempt to adjust tappet gap with engine running.

To adjust valve tappet gap on D-155 and D-179 engines. turn engine to position number one piston at top dead center of compression stroke. Adjust the four valves indicated on the chart shown in Fig. 49. Turn engine crankshaft one complete revolution to position number one piston at TDC of exhaust stroke and adjust the remaining two valves indicated on the chart.

To adjust valve tappet gap on D-206, D-239 and D-268 engines, turn engine to position number one pis-

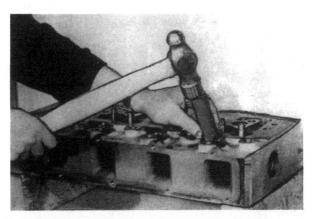

Fig. 48—Use installing tool 3055344R1 and drive nozzle sleeves into cylinder head until they bottom.

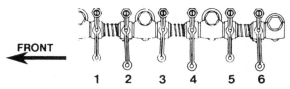

With	Adjust Valves (Engine Warm)					
No. 1 piston at T.D.C. (Compression)	1	2		4	5	
No. 1 piston at T.D.C. (Exhaust)			3			6

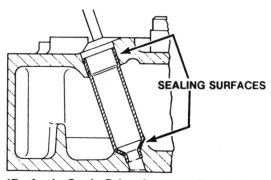

Fig. 47—Apply Grade B Loctite to sealing surfaces of nozzle sleeve.

SEALING SURFACES

FRONT

1 2 3 4 5 6

Fig. 49—Chart shows valve tappet adjusting procedure for D-155 and D-179 three cylinder engines. Refer to text.

ton at top dead center of compression stroke. Adjust the four valves indicated on the chart shown in Fig. 50. Turn engine one complete revolution to position number four piston at TDC of compression stroke and adjust the remaining four valves indicated on the chart.

NOTE: Valve arrangement is exhaust-intake-exhaust-intake and so on starting from front of cylinder head.

VALVE GUIDES AND SPRINGS

All Models

38. Intake and exhaust valve guides are interchangeable. Guides should be pressed into cylinder head using special tool 3055699R1 or equivalent, to set distance from top of guide to seat of spring recess to 1.102 inches (28 mm). After installation, guides must be reamed to an inside diameter of 0.3940-0.3945 inch (10.00-10.02 mm). Valve stem to valve guide diametral clearance should be 0.0016-0.0025 inch (0.041-0.064 mm) for intake valves and 0.0025-0.0033 inch (0.064-0.084 mm) for exhaust valves. Maximum allowable clearance in all guides is 0.006 inch (0.152 mm).

Intake and exhaust valve springs are also interchangeable. Springs should have a free length of 2.07-2.18 inches (52.58-55.37 mm) and should test 145.5-158.7 lbs. (647.2-705.90 N) when compressed to a length of 1.346 inch (34.19 mm). Renew any spring which is rusted, discolored or does not meet pressure test specifications.

VALVE TAPPETS
(CAM FOLLOWERS)

All Models

39. The 0.7862-0.7868 inch (19.97-19.98 mm) diameter mushroom type tappets operate directly in the unbushed crankcase bores. Clearance of tappets in bores should be 0.0004-0.0023 inch (0.010-0.058 mm).

Tappets can be removed after first removing the crankcase side plate and the camshaft. Oversize tappets are not available.

VALVE TAPPET LEVERS
(ROCKER ARMS)

All Models

40. Removal of rocker arms and shaft assembly is obvious after removal of rocker arm cover. To remove rocker arms from shaft, remove bracket clamp bolts and slide all parts from shaft. Outside diameter of shaft is 0.848-0.850 inch (21.54-21.59 mm). Rocker arms should have an operating clearance of 0.001-0.003 inch (0.025-0.076 mm) on rocker shaft with a maximum allowable clearance of 0.006 inch (0.152 mm).

Reassemble rocker arms on rocker shaft, keeping the following points in mind: Thrust washers are used between each spring and rocker arms. Spacer rings are used between rocker arms and brackets except between rear rocker arm and rear bracket. To ensure that lubrication holes in rocker shaft are in correct position, align punch mark on front end of rocker shaft with clamping slot in front mounting bracket as shown in Fig. 51. End of shaft must also be flush with bracket. Tighten rocker shaft clamp screws on brackets to a torque of 7.5-9.0 ft.-lbs. (10-12 N·m). Tighten rocker shaft assembly retaining cap screws to a torque of 44-52 ft.-lbs. (60-70 N·m).

TIMING GEAR COVER

All Models

41. **REMOVE AND REINSTALL.** To remove the timing gear cover, first disconnect battery cables and drain cooling system. Unbolt and remove hood and side panels. Identify and disconnect oil cooler lines, then remove oil cooler. Cap or plug all openings. On models so equipped, remove air conditioning con-

With	Adjust Valves (Engine Warm)							
No. 1 Piston at T.D.C. (Compression)	1	2		4	5			
No. 4 Piston at T.D.C. (Compression)			3			6	7	8

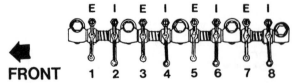

Fig. 50—Chart shows valve tappet adjusting procedure for D-206, D-239 and D-268 four cylinder engines. Refer to text.

Fig. 51—Align punch mark on end of rocker shaft with clamping slot in front mounting bracket on all models.

denser and tie condenser and hoses back out of the way. Unbolt and remove lower radiator seal strip. Remove wiring harness from fan shroud and lay rearward. Remove radiator side plate bolts, then unbolt and remove upper radiator air baffle. Remove upper and lower radiator hoses. Unbolt and remove radiator.

If so equipped, unbolt and remove front weights and weight bracket. Remove front axle assembly as in paragraph 4, 5 or 8 and front support as in paragraph 7 or 9. Identify and disconnect heater hoses if so equipped, at front of engine. Remove fan, fan belt and air conditioner compressor drive belt. Unbolt and remove the alternator and disconnect tachometer cable. Disconnect water manifold hose clamp. Remove fan spacer and water pump pulley. Unbolt and remove water pump and carrier assembly. Remove three cap screws and flat washer, tap crankshaft pulley with a plastic hammer to loosen clamping ring or pressure rings, then remove crankshaft pulley. Remove timing gear cover and oil pan front bolts, then pull cover forward off the dowels and remove from engine.

Use new gaskets and reinstall timing gear cover by reversing the removal procedure. Tighten cover retaining cap screws and water pump and carrier retaining cap screws to a torque of 19 ft.-lbs. (26 N·m). Tighten crankshaft pulley retaining cap screws in three steps, to a torque of 22 ft.-lbs. (30 N·m) in first step, 44 ft.-lbs. (60 N·m) in second step and 47 ft.-lbs. (65 N·m) in the final step.

TIMING GEARS

All Models

42. CRANKSHAFT GEAR. Crankshaft gear is a shrink fit on crankshaft. To renew the gear, it is recommended that crankshaft be removed from engine. Then, using a hammer and chisel, split the gear at its timing slot.

The roll pin for indexing crankshaft gear on crankshaft must protrude approximately 5/64 inch (2.0 mm). Heat new gear to 400° F (204° C) and install it against bearing journal.

> **WARNING: Always wear heat protective gloves when handling heated parts.**

Make certain all timing marks are aligned as shown in Fig. 52.

43. CAMSHAFT GEAR. Camshaft gear is a shrink fit on the camshaft. To renew the gear, remove camshaft as outlined in paragraph 46. Gear can now be pressed off in conventional manner, using care not to damage the tachometer drive slot in end of camshaft.

Before reinstalling, check thickness of thrust plate. If thrust plate thickness is less than 0.274 inch (6.96 mm), renew the plate.

When reassembling, install thrust plate and Woodruff key. Heat gear to 400° F (204° C) and install it on camshaft, setting thrust plate clearance to 0.004-0.018 inch (0.10-0.45 mm).

> **WARNING: Always wear heat protective gloves when handling heated parts.**

Install camshaft assembly and make certain all timing marks are aligned as shown in Fig. 52.

44. IDLER GEAR. To remove idler gear, first remove timing gear cover as outlined in paragraph 41. Idler gear shaft is attached to front of engine by a special (left hand thread) cap screw.

Idler gear is equipped with two renewable needle bearings. A spacer is used between the bearings.

When installing idler gear, coat threads of special cap screw with Loctite 41 and tighten cap screw to a torque of 72-76 ft.-lbs. (100-105 N·m). End clearance of gear on shaft should be 0.008-0.013 inch (0.20-0.33 mm). Align timing marks on all timing gears as shown in Fig. 52.

Timing gear backlash should be as follows:
Idler to crankshaft gear:
New gears .0.007-0.015 in.
(0.178-0.381 mm)
Used gears (max.). . . . . .0.027 in.
(0.686 mm)
Idler to camshaft gear:
New gears . . . . .0.0035-0.0106 in.
(0.089-0.269 mm)
Used gears (max.). . . . . .0.0226 in.
(0.574 mm)

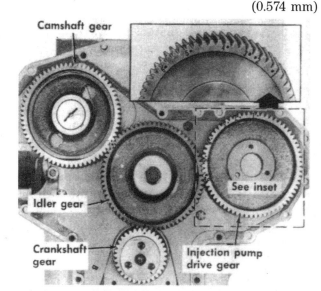

Fig. 52—Gear train and timing marks. On injection pump drive gear, use dot next to number 3 for D-155 and D-179 engines and dot next to number 4 for D-206, D-239 and D-268 engines.

Idler to injection pump drive gear:
New gears . . . . .0.002-0.012 in.
(0.051-0.305 mm)
Used gears (max.). . . . . .0.024 in.
(0.609 mm)

45. INJECTION PUMP DRIVE GEAR. To remove the injection pump drive gear, first remove timing gear cover as outlined in paragraph 41. Remove drive shaft nut and the three hub cap screws. Attach a puller to threaded holes in gear and pull gear and drive hub from shaft.

When reassembling, make certain all timing marks are aligned as shown in Fig. 52. The pump drive gear, which is used on all engines is equipped with several timing marks. Use dot next to number 3 for D-155 and D-179 engines and dot next to number 4 for D-206, D-239 and D-268 engines. Refer to paragraph 61 and retime injection pump. Tighten drive shaft nut to a torque of 44-50 ft.-lbs. (60-70 N·m) and the three hub cap screws to a torque of 19-21 ft.-lbs. (25-28 N·m).

CAMSHAFT AND BEARINGS

All Models

46. CAMSHAFT. To remove the camshaft, first remove the timing gear cover as outlined in paragraph 41. Then, remove rocker arm cover, rocker arms and shaft assembly and push rods. Unbolt and remove engine side cover.

Working through openings in camshaft gear, remove camshaft thrust plate retaining cap screws. Carefully withdraw camshaft and remove valve tappets (cam followers) as camshaft is being removed.

Press camshaft from camshaft gear, then remove camshaft thrust plate and measure the thickness. If thrust plate thickness is less than 0.274-0.276 inch (6.96-7.01 mm), renew thrust plate. Check diameter of camshaft journals which should be 2.2823-2.2835 inches (57.97-58.00 mm). If less than specifications, renew camshaft.

Refer to paragraph 43 and reinstall camshaft gear, camshaft thrust plate and camshaft.

47. CAMSHAFT BEARINGS. To remove camshaft bearings, first remove engine as outlined in paragraph 34 and camshaft as in paragraph 46. Unbolt and remove clutch, flywheel and engine rear end plate. Remove expansion plug behind camshaft rear bearing and remove the bearings.

NOTE: Camshaft bearings are furnished semifinished for service and must be align reamed after installation to an inside diameter of 2.2844-2.2856 inches (58.024-58.054 mm).

Install new bearings so that oil holes in bearing are in register with oil holes in crankcase. Normal operating clearance of camshaft journals in bearings is 0.0009-0.0033 inches (0.023-0.084 mm). Maximum allowable clearance is 0.0045 inch (0.115 mm).

When installing expansion plug at rear camshaft bearing, apply a light coat of sealing compound to edge of plug and bore.

ROD AND PISTON UNITS

All Models

48. Connecting rod and piston assemblies can be removed from above after removing the cylinder head, oil pan, oil pump pressure and suction pipes and engine balancer on models so equipped.

Cylinder numbers are stamped on the connecting rod and cap. Stamp any new or unmarked rod and cap assemblies with correct cylinder numbers. Numbers on rod and cap must be in register and face toward camshaft side of engine. The arrow or FRONT marking stamped on tops of pistons must be toward front of engine.

Tighten the connecting rod nuts to a torque of 18.5 ft.-lbs. (25 N·m), then tighten nuts an additional 120 degrees (two flats on nut).

PISTONS, SLEEVES AND RINGS

All Models

49. Pistons are not available as individual service parts, but only as matched units with wet type sleeves. New pistons have a diametral clearance in new sleeves of 0.0035-0.005 inch (0.09-0.13 mm) when measured between piston skirt and sleeve at 90 degrees to piston pin.

The wet type cylinder sleeves should be renewed when out-of-round or taper exceeds 0.006 inch (0.152 mm). Inside diameter of new sleeve is 3.875-3.8759 inches (98.425-98.448 mm) on D-155, D-179, D-206 and D-239 engines or 3.937-3.9379 inches (100.0-100.023 mm) on D-268 engines. Cylinder sleeves are removed, using correct adapter plate and sleeve removing tool.

Sleeves are plateau honed and should never be deglazed by honing on rering jobs. Remove oil and varnish glaze with a 50/50 mixture of hot water and liquid detergent.

Before installing new sleeves, thoroughly clean counterbore at top and seal ring groove at bottom. All sleeves should enter crankcase bores full depth and should be free to rotate by hand when tried in bores without seal rings. After making a trial installation without seal rings, remove sleeves and install new seal rings dry into grooves in crankcase.

NOTE: Two different seal rings have been used. Measure groove diameter to determine the correct seal ring to be installed. If groove diameter measures 4.667-4.677 inches (118.54-118.80 mm), install old type (red tracers). If groove diameter measures 4.658-4.668 inches (118.32-118.58 mm), install new type (gold tracers).

Coat lower end of sleeve with a light coat of clean motor oil and install sleeve. If seal ring is in place and not pinched, very little hand pressure is required to push sleeve completely in place.

Sleeve flange should extend 0.0031-0.0047 inch (0.079-0.119 mm) above top surface of cylinder block. If sleeve stand-out is excessive, check for foreign material and under sleeve flange. If sleeve stand-out is less than 0.0031 inch (0.079 mm), install shim under sleeve flange as necessary. Shims are available in thicknesses of 0.002, 0.004, 0.010, 0.020 and 0.031 inch (0.05, 0.10, 0.25, 0.50 and 0.80 mm).

Pistons are fitted with two compression rings and one oil ring. The top compression ring is a chrome faced rectangular ring and upper side is marked TOP.

The second compression ring is a taper face ring and is installed with largest outside diameter toward bottom of piston. Upper side of ring is marked TOP.

The oil control ring can be installed either side up, but make sure the coil spring expander is completely in its groove.

Additional piston ring information is as follows:

Ring End Gap:
Compression rings . . . . .0.014-0.022 in.
(0.035-0.056 mm)
Oil control ring . . . . .0.010-0.016 in.
(0.025-0.040 mm)
Ring Side Clearance:
Top compression ring . . . . .0.0035-0.0048 in.
(0.089-0.122 mm)
Second compression ring . . . . . .0.0030-0.0042 in.
(0.076-0.107 mm)
Oil control ring . . . . .0.0014-0.0024 in.
(0.035-0.061 mm)

PISTON PINS

All Models

50. The full floating type piston pins are retained in the piston bosses by snap rings. Specifications are as follows:

Piston pin diameter . . . . .1.4172-1.4174 in.
(35.997-36.002 mm)
Piston pin diametral
clearance in piston . . . . . .0.0000-0.0001 in. tight
(0.0001-0.0025 mm tight)
Maximum allowable . . . . .0.0002 in. loose
(0.0051 mm loose)

Piston pin diametral
clearance in bushing . . . . .0.0005-0.0011 in.
(0.0127-0.0279 mm)
Maximum allowable . . . . .0.002 in.
(0.0508 mm)

Piston pin bushings are serviced semi-finished and must be reamed or honed for correct pin fit after they are pressed into connecting rods. Oversize piston pins are not available.

CONNECTING RODS AND BEARINGS

All Models

51. Connecting rod bearings are of the slip-in, precision type, renewable from below after removing oil pan, oil pump pressure and suction pipes, engine balancer (if so equipped) and connecting rod caps.

When installing new bearing inserts, make certain that projections on same engage slots in connecting rod and cap. Also, make certain that cylinder identifying numbers on rod and cap are in register and face toward camshaft side of engine. Numbers on rod and cap should be stamped on same side as the flat boss on small end of rod.

Connecting rod bearings are available in standard size as well as undersizes of 0.010, 0.020 and 0.030 inch (0.25, 0.51 and 0.76 mm).

Check crankshaft crankpins and connecting rod bearings against the values which follow:

Crankpin diameter . . . . .2.5185-2.5193 in.
(63.97-63.99 mm)
Max. allowable out-of-round . . . . .0.0015 in.
(0.038 mm)
Max. allowable taper
per inch (25.4 mm) . . . . .0.0015 in.
(0.038 mm)
Rod bearing diametral clearance . . .0.002-0.004 in.
(0.051-0.102 mm)
Rod side clearance . . . . .0.006-0.012 in.
(0.152-0.305 mm)

Tighten connecting rod nuts to a torque of 18.5 ft.-lbs. (25 N·m), then tighten nuts an additional 120 degrees (two flats on nut).

CRANKSHAFT AND MAIN BEARINGS

All Models

52. Crankshaft is supported in four main bearings on D-155 and D-179 engines and five main bearings on D-206, D-239 and D-268 engines. Main bearings are of the nonadjustable, slip-in, precision type, renewable from below after removing clutch and flywheel for access to the two cap screws securing the seal bar below rear main bearing cap and then, removing oil pan, oil pump pressure and suction pipes, engine balancer (if so equipped), seal bar, timing gear cover on four cylinder models and main bearing caps. Crankshaft end play is controlled by the flanged rear main bearing inserts. Removal of crankshaft requires removal of timing gear cover on all models.

Check crankshaft and main bearings against the values which follow:

Crankpin diameter . . . . .2.5185-2.5193 in.
(63.97-63.99 mm)
Main journal diameter . . . . .3.1484-3.1492 in.
(79.97-79.99 mm)
Max. allowable out-of-round . . . . .0.002 in.
(0.051 mm)
Crankshaft end play . . . . .0.006-0.009 in.
(0.152-0.229 mm)
Main bearing diametral
clearance . . . . .0.0028-0.0055 in.
(0.071-0.140 mm)

Tighten main bearing bolts in three stages; 30 ft.-lbs. (40 N·m), 60 ft.-lbs. (80 N·m) and 110 ft.-lbs. (150 N·m). Main bearings are available in standard size as well as undersizes of 0.010, 0.020 and 0.030 inch (0.25, 0.51 and 0.76 mm).

CRANKSHAFT SEALS

All Models

53. FRONT. To renew the crankshaft front oil seal, first disconnect battery cables and drain cooling system. Unbolt and remove hood, grille and side panels. Identify and disconnect oil cooler lines, then remove oil cooler. Cap or plug all openings. On models so equipped, remove air conditioning condenser and tie condenser and hoses back out of the way. Unbolt and remove lower radiator seal strip. Remove wiring harness from fan shroud and lay rearward. Remove radiator side plate bolts, then unbolt and remove upper radiator air baffle. Remove upper and lower radiator hoses. Unbolt and remove radiator.

If so equipped, unbolt and remove front weights and weight bracket. Remove front axle assembly as outlined in paragraph 4, 5 or 8 and front support as in paragraph 7 or 9. Remove fan and alternator belt. Remove three cap screws and flat washer, tap crankshaft pulley with a plastic hammer to loosen clamping ring or pressure rings, then remove crankshaft pulley. Remove spacer ring and "O" ring from crankshaft, then remove oil seal.

Inspect spacer ring and timing pin for wear or other damage and renew as necessary. Use a nonhardening sealer on timing pin. When installing spacer ring, timing pin must engage slot in crankshaft gear. Us-

ing special tool No. 3228341R1 or equivalent, install new seal with lip toward inside. When installing crankshaft pulley, align pulley with timing pin. Make certain clamping ring or pressure rings are properly positioned, then install flat washer and three cap screws. Tighten the cap screws evenly in three steps to a torque of 22 ft.-lbs. (30 N·m) in first step, 44 ft.-lbs. (60 N·m) in second step and 47 ft.-lbs. (65 N·m) in the final step.

Reassembly is the reverse of disassembly procedures.

54. REAR. To renew crankshaft rear oil seal, first apply park brake and securely block rear wheels. Disconnect battery cables and remove battery. Remove hood, grille and side panels. Disconnect tachometer cable, starter wiring and rear wiring connector. Shut off fuel and disconnect supply line and fuel return line. Cap all openings. Disconnect injection pump speed control and shut off cables. Identify and disconnect oil cooler lines and steering lines and cap all openings. On models so equipped, disconnect heater hoses and air conditioning lines. On models equipped with front drive axle, remove drive shaft shield and front drive shaft. If available, attach a split stand under clutch housing and engine oil pan or block up under clutch housing and attach a hoist to engine. Place wood blocks between front axle and front support to prevent tipping. Unbolt clutch housing from engine and move front half of tractor forward.

Unbolt and remove engine clutch and flywheel. Remove seal housing retaining cap screws, then remove seal housing and rear oil seal. Remove seal from housing.

Inspect sealing face on crankshaft for wear groove. New seal can be installed in any of three locations: 0.060 inch (1.5 mm) above flush with rear face of housing (new engine original position), flush with housing or 0.060 (1.5 mm) below flush with housing. Location of seal will depend on condition of sealing surface on crankshaft. Use new gasket and install seal housing with new oil seal. Tighten oil seal housing cap screws to a torque of 11 ft.-lbs. (15 N·m), then retighten to a torque of 14 ft.-lbs. (19 N·m). Reassemble by reversing disassembly procedure.

FLYWHEEL

All Models

55. To remove the flywheel, refer to procedures outlined in paragraph 54. To install new flywheel ring gear, heat same to approximately 400-520° F (200-270° C).

WARNING: Always wear heat protective gloves when handling heated parts.

Inspect face of flywheel and if face is grooved more than 0.025 inch (0.64 mm) or if face is worn out-of-flat more than 0.006 inch (0.152 mm), renew or reface flywheel. A maximum of 0.125 inch (3.17 mm) can be removed from flywheel face.

NOTE: The same amount of material removed from flywheel face must be removed from pressure plate mounting surface on flywheel.

Tighten flywheel retaining cap screws in three steps: 30 ft.-lbs. (40 N·m) in the first step, 60 ft.-lbs. (80 (N·m) in the second step and 110 ft.-lbs. (150 N·m) in the final step.

ENGINE BALANCER

Models So Equipped

56. R&R AND OVERHAUL. All D-206, D-239 and D-268 engines are equipped with an engine balancer which is mounted on underside of engine crankcase. Balancers are driven by a renewable gear (2—Fig. 53) which is a shrink fit on crankshaft. The balancer consists of two unbalanced gear weights which rotate in opposite directions at twice crankshaft speed. They produce forces which tend to counteract the vibration which is inherent in four cylinder engines having a single plane crankshaft (1 and 4 throws displaced 180 degrees from throws 2 and 3). It is

extremely important that balancer weights are correctly timed to each other and that complete unit is timed to crankshaft.

To remove the engine balancer, drain oil and remove oil pan. Unbolt and remove oil pump pressure and suction pipe assembly. Unbolt and remove balancer from crankcase.

To disassemble the balancer unit, drive roll pins (8) out of housing and into weight shafts (5). Press or drive shafts out roll pin side of housing (7) and lift out weight gear assemblies (3 and 4). Drive roll pins (8) out of shafts.

Clean and inspect all parts for excessive wear or other damage. Refer to specifications which follow to determine parts renewal and operating clearances.

Backlash, crankshaft gear to weight gear:
D-206 and D-239 engines . . . . 0.010-0.016 in.
(0.25-0.40 mm)
D-268 engine . . . . 0.009-0.020 in.
(0.23-0.51 mm)
Backlash between weight gears . . . . 0.007-0.009 in.
(0.18-0.23 mm)
Weight gear bushing ID . . . . 0.9082-0.9094 in.
(23.068-23.098 mm)
Shaft bearing surface diameter . . 0.9051-0.9055 in.
(22.989-22.999 mm)
Weight gear operating clearance
on shaft . . . . 0.0027-0.0043 in.
(0.069-0.109 mm)
Weight gear end clearance
in housing . . . . 0.008-0.016 in.
(0.203-0.406 mm)

NOTE: Weight gears are serviced only as a matched set. If bushings show excessive wear, renew both weight gear assemblies.

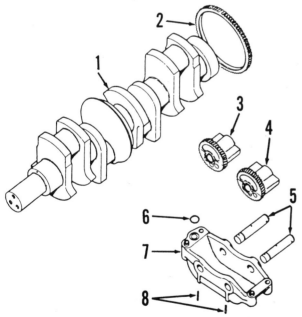

Fig. 53—Exploded view of engine balancer used on D-206, D-239 and D-268 four cylinder engines. Balancer drive gear (2) is a shrink fit on crankshaft.

1. Crankshaft
2. Balancer drive gear
3. Driven weight assy.
4. Drive weight assy.
5. Weight shafts
6. "O" ring
7. Housing
8. Double roll pins

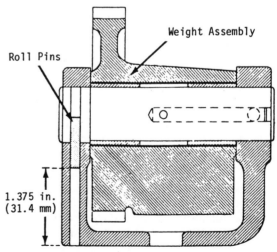

Fig. 54—Sectional view of balancer weight gear, housing and shaft showing correct installation of roll pins. Roll pins must be shorter than shaft diameter.

When reassembling, make certain that oil passages in housing and shafts are clean. Lubricate shafts and weight gear bushings and place drive weight gear (4) in housing. Install weight shaft from roll pin side of housing through the weight gear. Align roll pin hole in shaft with roll pin hole in housing, then tap shaft into position. Install double roll pins (8) as shown in Fig. 54.

NOTE: Roll pins must be shorter than shaft diameter.

Install roll pins to a depth of 1.375 inches (31.4 mm) in roll pin bosses. Place idler weight in housing and with both weights down, mesh teeth so that timing marks are aligned. Install second shaft and secure with double roll pin. Using a dial indicator, check backlash between weight gears. Backlash should be 0.007-0.009 inch (0.18-0.23 mm). If backlash is excessive, renew weight gear assemblies.

To install balancer assembly, first rotate crankshaft to position number two piston at TDC of compression stroke. Place new "O" ring (6—Fig. 53) in recess on balancer. Then, with weights toward bottom of balancer housing and weight gear timing marks aligned, install balancer so that timing mark on balancer drive weight is aligned with timing mark on balancer drive gear (2). Use new "O" rings and install oil pump pressure and suction pipe assembly. Using a dial indicator, check backlash between balancer drive weight gear (4) and balancer drive gear (2). Backlash should be 0.010-0.016 inch (0.25-0.40 mm) on D-206 and D-239 engines or 0.009-0.020 inch (0.23-0.51 mm) on D-268 engine.

NOTE: When backlash between weight gears is satisfactory and backlash between weight gear (4) and balancer drive gear (2) is excessive, drive gear (2) is excessively worn and should be renewed.

Install oil pan with new pan gasket and fill crankcase to proper level with new oil.

57. RENEW BALANCER DRIVE GEAR. To renew balancer drive gear (2—Fig. 53), first remove crankshaft from engine. Remove worn or damaged drive gear by using a chisel and hammer and splitting the gear. Remove any burrs which might be present on gear mounting surface of crankshaft. Heat new gear to 360-390° F (180-200° C), align the single timing mark (below tooth space of drive gear) with the notch on crankshaft flange and slide drive gear on crankshaft. The two marked teeth on drive gear should be toward flywheel end of crankshaft.

WARNING: Always wear heat protective gloves when handling heated parts.

OIL PUMP AND RELIEF VALVE

All Models

58. The internally mounted gear type oil pump is driven from the crankshaft gear and is accessible after removing the oil pan on all models and the timing gear cover on D-206, D-239 and D-268 engines. The pump is mounted to the front main bearing cap on all models.

On D-155 and D-179 engines, remove the main bearing cap and oil pump as an assembly. Leave bearing cap on pump when repairing oil pump.

On D-206, D-239 and D-268 engines, use an Allen wrench and remove the two upper pump retaining bolts, then working through holes in idler gear, remove the remaining two retaining bolts. Unbolt pressure and suction pipe assembly, then remove oil pump assembly.

To disassemble the oil pump, remove pressure and suction pipe assembly (16—Fig. 55) and "O" rings (15 and 21). Remove nut (3) and pull idler gear (8) and bearings from front cover (10). Remove bearings (4 and 7) washer (5) and snap ring (6) from idler gear. Remove nuts from rear of body bolts (1) and remove rear cover (14). Remove body (11) and driven pump

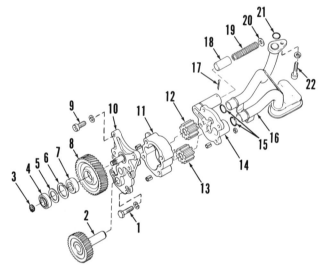

Fig. 55—Exploded view of the oil pump used on all models.

1. Body bolt (4)
2. Drive gear & shaft
3. Nut
4. Tapered roller bearing assy.
5. Washer
6. Snap ring
7. Tapered roller bearing assy.
8. Idler gear
9. Socket head cap screw (4)
10. Front cover
11. Body
12. Driven pump gear
13. Drive pump gear
14. Rear cover
15. "O" rings
16. Pressure & suction pipes
17. Cotter pin
18. Relief valve
19. Spring
20. Washer
21. "O" ring
22. Socket head cap screw

gear (12). Pump gears (12 and 13) are serviced as a matched set and if either gear is faulty, renew both gears. Gear (13) is a shrink fit on shaft (2) and if faulty, split the gear to remove. Withdraw drive gear and shaft (2).

Remove cotter pin (17), then remove washer (20), spring (19) and relief valve (18). Specifications for relief valve are as follows:

Valve diameter . . . . .0.825-0.827 in.
(20.95-21.01 mm)
Valve clearance in bore. . . . . .0.003-0.007 in.
(0.076-0.178 mm)
Valve spring (used):
Free length . . . . .2.52 in.
(64.00 mm)
Test length . . . . .1.858 in.
(47.19 mm)
Test load . . . . .18.0-20.0 lbs.
(80.06-88.96 N)
Valve spring (new):
Free length . . . . .2.85 in.
(72.39 mm)
Test length . . . . .1.86 in.
(47.24 mm)
Test load . . . . .33.0 lbs.
(146.78 N)

Inspect all parts for scoring, excessive wear or other damage. Check pump against the following specifications:

Drive shaft end play (cover
installed) .0.000-0.002 in.
(0.000-0.051 mm)
Drive shaft running clearance
in bushings . . . . .0.001-0.003 in.
(0.025-0.076 mm)
Radial clearance (gears
to body) . . . . .0.007-0.012 in.
(0.178-0.305 mm)

Idler gear end play . . . . .0.000-0.002 in.
(0.000-0.051 mm)
Oil pressure at rated rpm and
194-212° F (90-100° C) . . . . .40-60 psi
(275-414 kPa)

When installing gear (13), heat gear in a bearing oven to a temperature of 395-750° F (200-400° C) and slide on drive shaft until tight against a 0.004 inch (0.1016 mm) feeler gage.

WARNING: Always wear heat protective gloves when handling heated parts.

Tighten body bolts (1) evenly using a crosswise sequence. Install idler gear (8) with short hub side toward oil pump. Using a new nut (3), tighten nut until idler gear end play is 0.000-0.002 inch (0.000-0.051 mm).

Reinstall oil pump assembly. Use new "O" rings (15 and 21) when installing oil pressure and suction pipe assembly. Balance of reassembly is the reverse of disassembly.

OIL PAN

All Models

59. REMOVE AND REINSTALL. To remove the oil pan, place a jack under the clutch housing. Remove drain plug and drain oil. Remove the lower four bolts and loosen the upper two bolts in the front support to engine. Unbolt oil pan from clutch housing. Place a floor jack under oil pan, remove pan bolts and lower oil pan from tractor.

Use new gasket and reinstall oil pan. Tighten pan bolts evenly to a torque of 24 ft.-lbs. (32 N·m). Tighten clutch housing bolts to a torque of 80 ft.-lbs. (108 N·m) and front support bolts to a torque of 220 ft.-lbs. (298 N·m). Fill crankcase to proper level with new oil.

DIESEL FUEL SYSTEM

The diesel fuel system on all models is the direct injection type equipped with a Robert Bosch injection pump.

When servicing any unit of the diesel fuel system, the maintenance of absolute cleanliness is of utmost importance. Of equal importance is the avoidance of nicks or burrs on any of the working parts.

Probably the most important precaution that service personnel can impart to owners of diesel powered tractors, is to urge them to use an approved fuel that is absolutely clean and free from foreign material. Extra precaution should be taken to make certain that no water enters the fuel storage tank.

FILTERS AND BLEEDING

All Models

60. All models are equipped with two diesel fuel filters. On Models 385 and 485, the primary filter is located on right side of engine and the secondary filter is on the left side. See Fig. 56. On Models 585, 685 and 885, both filters are on left side of engine (Fig. 57). To renew filter elements, first shut off fuel valve on models without cab. Then, on all models, open air vent screws at each filter head. Open drain valves and allow fuel to drain from filters. Using a strap wrench or filter wrench, unscrew and remove primary and secondary filters.

When installing new filters, use new seal rings and apply a light coat of oil to seal ring. Screw filters on until seal ring contacts filter head, then hand tighten 1/2 to 3/4 turn. Close drain valve.

On all models without cab, diesel fuel is gravity fed to filters. All models equipped with cab are equipped with a hand primer pump. On models without cab, open shut-off valve under right side of fuel tank. On all models, open air vent screw on primary filter head. On models with cab, actuate hand primer

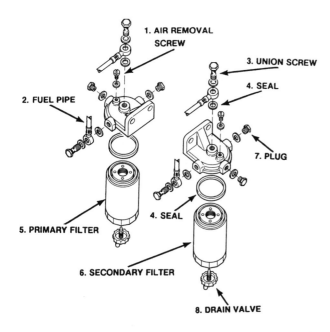

FUEL FILTERS 3 CYLINDER ENGINE

Fig. 56—Exploded view of fuel filters used on three cylinder engines. Primary filter is on right side of engine and secondary filter is on left side.

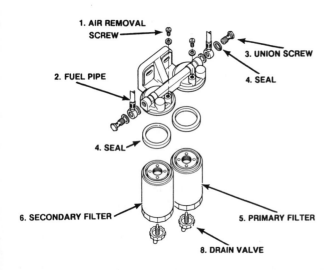

FUEL FILTERS 4 CYLINDER ENGINE

Fig. 57—Exploded view of fuel filters used on four cylinder engines. Filters are on left side of engine.

pump. On all models, close air vent screw when bubble free fuel flows from air vent. Bleed air from secondary filter in the same manner. Loosen inlet line at injection pump and bleed air at this point. Tighten the banjo bolt when fuel runs a solid stream with no air bubbles.

NOTE: The life of filter elements depends on the amount of contamination they must remove from the fuel. If engine misfires or loss of power is evident, renew the elements. In any event, filter elements must be renewed after each 800 hours of operation.

INJECTION PUMP

All models are equipped with a Robert Bosch Model VE diesel injection pump which is a distributor type and is completely sealed.

Because of the special equipment needed, and skill required of servicing personnel, service of injection pumps is generally beyond the scope of that which should be attempted in the average shop. Therefore, this section will include only timing of pump to engine, removal and installation and the linkage adjustments which control engine speeds.

If additional service is required, the pump should be turned over to a Case-International facility which is equipped for diesel service, or to some other authorized diesel service station. Inexperienced personnel should NEVER attempt to service diesel injection pumps.

All Models

61. R&R AND SET STATIC TIMING. To remove the injection pump, first thoroughly clean the pump, lines and side of engine. On models without cab, shut off fuel valve under right side of fuel tank. Then, on all models, disconnect shut-off and throttle cables and/or rod. Remove hood, grille and left side panel. Remove air cleaner assembly, then unbolt and remove rocker arm cover. Remove clamp from injector lines and disconnect the injector lines from pump. Disconnect fuel inlet and excess fuel return lines from pump. Cap or plug all line openings immediately. Rotate engine crankshaft in normal direction of rotation until timing pointer is aligned with the "TC" (top dead center) mark on crankshaft pulley. At this time, both rocker arms for No. 1 cylinder should be loose. If so, engine is at TDC of compression stroke of No. 1 cylinder. If not, rotate engine crankshaft one complete revolution until timing pointer is aligned with the "TC" mark on pulley. This will position No. 1 piston at TDC of compression stroke. Then, turn crankshaft pulley counterclockwise (viewed from front) until the 8 degree BTDC mark on pulley is aligned with timing pointer. Remove the two nuts and washers from the injection pump mounting studs.

Unbolt and remove pump drive gear access cover from timing gear cover. Remove nut and lockwasher from end of pump drive shaft, then remove the three cap screws and washers from the drive hub. Thread the three bolts of a hub extractor or other similar puller into holes in drive gear, then tighten center bolt against end of pump shaft until hub is free of shaft. Carefully remove injection pump, taking care not to lose the key from pump shaft. Remove puller and drive hub.

CAUTION: DO NOT rotate crankshaft after injection pump is removed.

When reinstalling, place a new key in drive shaft, then rotate drive shaft to align key with No. 1 injector outlet on pump as shown in Fig. 58. Apply a coat of Loctite silicone sealant to both sides of a new mounting gasket and install gasket over pump mounting studs. Install injection pump, taking care not to lose the key. Install washers and nuts and tighten nuts to a torque of 16.0-18.0 ft.-lbs. (22.0-25.0 N·m). Make certain that mark on pump mounting flange is aligned with mark on engine front plate. Align slot in drive hub with key in drive shaft and install drive hub. Install the three cap screws and washers and secure drive hub to drive gear. Install lockwasher and nut on end of pump shaft, then tighten nut to a torque of 44-50 ft.-lbs. (60-70 N·m). Check to be sure that the 8 degree BTDC timing mark on crankshaft pulley is aligned with timing pointer. Remove the plug from center of pump distributor head. Install a special dial probe extension CAS-1745 in distributor head. Install a dial indicator probe into the extension until a reading of 0.200 inch (5.0 mm) is shown on the dial. Tighten collar on extension to hold dial indicator in this position. Loosen the three hub cap screws. Turn the injection pump drive shaft counter-

clockwise (viewed from front), until indicator needle stops moving. At this time, plunger in pump is bottomed. Zero the dial indicator, then turn pump drive shaft clockwise (viewed from front), until dial indicator shows a reading of 0.039 inch (1.0 mm). Evenly tighten the three hub cap screws to a torque of 19-21 ft.-lbs. (25-28 N·m). Turn crankshaft pulley counterclockwise (viewed from front), until dial indicator shows zero reading. Then, turn crankshaft pulley clockwise (viewed from front), until dial indicator reads 0.039 inch (1.0 mm). At this time, if the 8 degree BTDC timing mark on crankshaft pulley is aligned with timing pointer, injection pump is in time.

To check pump plunger, continue turning crankshaft pulley clockwise until dial indicator shows a maximum reading (plunger at end of outward travel). Reading should be 0.098-0.099 inch (2.49-2.51 mm) on Models 385 and 885, 0.086-0.087 inch (2.19-2.21 mm) on Model 485 and 0.110-0.111 inch (2.79-2.81 mm) on Models 585 and 685.

Remove dial indicator and probe extension, then install plug with new copper washer. Tighten plug to a torque of 10-15 ft.-lbs. (14-20 N·m). Install drive gear access cover with new gasket. Tighten retaining stud nuts to a torque of 6-7 ft.-lbs. (8-9 N·m). Reconnect fuel inlet line, excess fuel return line and injector lines. Bleed air from fuel system as outlined in paragraph 60.

Connect pump shut-off cable and pump throttle cable or rod and adjust as outlined in paragraph 62. Complete reassembly by reversing disassembly procedures.

INJECTION PUMP CONTROL ADJUSTMENTS

All Models

62. LOW AND HIGH IDLE. To adjust the low idle speed, start engine and bring to operating temperature. Move throttle lever to low idle position and make certain that speed control lever (13—Fig. 59) is against low speed screw (6). If not, adjust linkage rod (models without cab) or cable (models with cab) as necessary. Then, loosen locknut and adjust screw (6) until idle speed is 725-750 rpm. Tighten locknut.

To adjust high idle speed, start engine and allow it to warm up to operating temperature. Move throttle lever to high idle position and make certain that speed control lever (13) is against high speed adjustment screw (8). If not, adjust linkage rod or cable as necessary. Then, loosen locknut and adjust (8) to obtain a high idle speed of 2430 rpm for Model 385, 2480 rpm for Model 485, 2590 rpm for Model 585, 2705 rpm for Model 685 and 2650 rpm for Model 885. Tighten locknut.

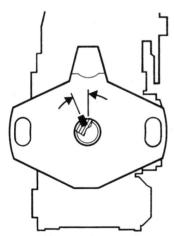

Fig. 58—Turn pump drive shaft so that key is aligned with No. 1 injector outlet of pump (first outlet counterclockwise from vertical, as viewed from drive shaft end of pump).

63. SHUT-OFF CONTROL. To adjust the shut-off control, pull shut-off knob fully outward. Then, check to see that lever (14—Fig. 59) contacts stop. If not, adjust cable position as required. Push knob fully inward and check to see that lever (14) contacts the pump case. If not, loosen locknut under knob, back off knob slightly and retighten locknut. Recheck to see that shut-off lever (14) contacts stops when knob is fully in run position or fully out in stop position.

INJECTION NOZZLES

WARNING: Fuel leaves the injection nozzles with sufficient force to penetrate the skin. When testing, keep your person clear of the nozzle spray.

All Models

64. TESTING AND LOCATING FAULTY NOZZLE. If engine does not run properly and a faulty injection nozzle is suspected, or if one cylinder is misfiring, locate faulty nozzle as follows: With engine running at low isle speed, loosen the high pressure line fitting on each injector in turn, thereby allowing fuel to escape at the fitting rather than enter the cylinder. As in checking spark plugs in a spark ignition engine, the faulty nozzle is the one that when

its line is loosened, least affects the running of the engine.

Remove the suspected nozzle as outlined in paragraph 65, place nozzle in a test stand and check nozzle against the following specifications:

Opening pressure (new) . . . . .3260-3380 psi
(22500-23300 kPa)
Opening pressure (used) . . . . .3050 psi
(21030 kPa)

Nozzle showing visible wetting on the tip after 10 seconds at 300 psi (2068 kPa) below opening pressure is permissible. Maximum allowable leakage through return is 10 drops in one minute at 300 psi (2068 kPa) below opening pressure. If nozzle requires overhauling, refer to paragraph 66.

65. REMOVE AND REINSTALL. To remove any injection nozzle, first remove hood, then clean nozzle, injection line, return line and cylinder head. Disconnect leakage return line and high pressure injection line from nozzle. Cap or plug all openings.

CAUTION: It is recommended that cooling system be drained before removing injectors. It is possible that injector nozzle sleeve may come out with injector and allow coolant to enter engine.

Remove clamp bolt and carefully withdraw injector assembly from cylinder head.

When reinstalling, use new "O" ring on injector and tighten clamp retaining bolt to a torque of 17-20 ft.-lbs. (23-27 N·m). Tighten injection line fitting to a torque of 36-51 ft.-lbs. (50-70 N·m).

66. OVERHAUL. To disassemble the injector, clamp flats of nozzle body (3—Fig. 60) in a vise with nozzle tip pointing upward. Remove nozzle holder nut (10). Remove nozzle tip (9) with valve (8) and spacer (7). Invert nozzle body (3) and remove spring seat (6), spring (5) and shim (4).

Thoroughly clean all parts in a suitable solvent. Clean inside the orifice end of nozzle tip with a wooden cleaning stick. Spray hole diameter is 0.0106 inch (0.27 mm) for Models 385, 485 and 585 and 0.0114 inch (0.29 mm) for Models 685 and 885. Clean spray holes with a cleaning wire that is 0.0005-0.001 inch (0.0127-0.0254 mm) smaller than diameter of spray holes. Cleaning wire must extend 0.0312 inch (0.79 mm) past tip of pin vise.

When reassembling, make certain all parts are perfectly clean and install parts while wet with clean diesel fuel. To check cleanliness and fit of valve (8) in nozzle tip (9), use a twisting motion and pull valve about 1/3 of its length out of nozzle tip. When released, valve should slide back to its seat by its own weight.

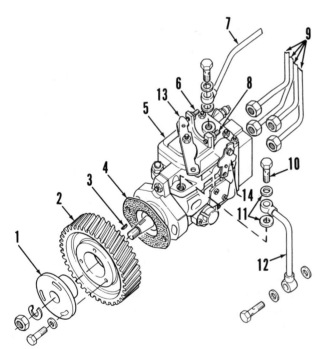

Fig. 59—View of typical fuel injection pump and relative components.

1. Drive hub	8. High speed screw
2. Drive gear	9. Injector lines
3. Key	10. Banjo bolt
4. Gasket	11. Seal rings
5. Injection pump	12. Fuel inlet line
6. Low speed screw	13. Speed control
7. Excess fuel	lever
return line	14. Shut-off lever

NOTE: Valve and nozzle tip are mated parts and under no circumstance should valves and nozzle tips be interchanged.

Install shim (4), spring (5) and spring seat (6) in nozzle body (3). Place spacer (7) and nozzle tip (9) with

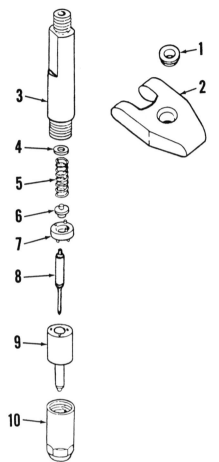

Fig. 60—Exploded view of Robert Bosch fuel injector nozzle assembly used on all models.

1. Washer	6. Spring seat
2. Clamp	7. Spacer
3. Nozzle body	8. Valve
4. Shim	9. Nozzle tip
5. Spring	10. Nozzle holder nut

valve (8) in position and install holder nut (10). Tighten holder nut to a torque of 40-50 ft.-lbs. (55-70 N·m).

Connect assembled injector nozzle to a test pump and check opening pressure. Adjust opening pressure by varying number and thickness of shims (4). Opening pressure should be adjusted to 3050 psi (21030 kPa) if old spring (5) is used or to 3260-3380 psi (22500-23300 kPa) if new spring is installed. Valve should not show leakage at orifice spray holes for 10 seconds at 300 psi (2068 kPa) below opening pressure. Maximum allowable leakage through return port is 10 drops in one minute at 300 psi (2068 kPa) below opening pressure.

ETHER STARTING AID

All Models

67. At temperatures below freezing, it is necessary that ether be used as a starting aid. To test the ether spray pattern, disconnect ether line at spray nozzle and remove spray nozzle from intake manifold. Reconnect spray nozzle to ether line. Press ether injection button and observe spray pattern. A good spray pattern is cone shaped. Dribbling or no spray indicates a blocked spray nozzle or lack of ether pressure. Clean spray nozzle or install new can of ether as required.

To change the ether fluid container, turn knurled nut at bottom of bail until container can be removed. Install new container in the bail and tighten the retaining nut while guiding container head into position. Rotate container to be sure it is seated properly in injector body, then tighten nut to hold container firmly in position.

NOTE: In warm temperatures, ether container may be removed and a protective plug installed in injector body. DO NOT operate tractor without either the ether container or protective plug in position.

COOLING SYSTEM

RADIATOR

All Models

68. To remove the radiator, first disconnect battery cables and drain cooling system. Remove hood, grille and front side panels. Identify and disconnect oil cooler lines and remove oil cooler. Cap or plug all openings. On models so equipped, remove air conditioning condenser and tie condenser and hoses back out of the way. Unbolt and remove lower radiator seal strip. Remove wiring harness from fan shroud and lay rearward. Remove radiator side plate bolts, then unbolt and remove upper radiator air baffle and air intake tube. Remove upper and lower radiator hoses. Unbolt and remove radiator assembly.

Reinstall by reversing removal procedure. Refill cooling system. Capacity is approximately 3.0 U.S. gallons (11.4 L) for Models 385 and 485 and 3.6 U.S. gallons (13.7 L) for Models 585, 685 and 885.

A 9-12 psi (62-82 kPa) pressure radiator cap is used on all models.

FAN

All Models

69. The five blade (6 blade with air conditioning) 18 inch (457 mm) diameter fan is attached to the water pump shaft. To remove the fan, first remove hood and front side panels. Drain the cooling system and disconnect upper end of lower radiator hose. Loosen adjusting bolts on alternator and loosen the belt. Unbolt fan shroud and move shroud rearward. Then, unbolt and remove fan.

Reinstall by reversing the removal procedure. Tighten fan mounting bolts to a torque of 8.5 ft.-lbs. (11.5 N·m). Adjust alternator-fan belt until a pressure of 25 lbs. (111 N), applied midway between fan pulley and crankshaft pulley, will deflect belt 3/4 inch (19 mm).

THERMOSTAT

All Models

70. All models are equipped with a single thermostat located under the water outlet (at rear of upper radiator hose). Thermostat should start to open at 180° F (84° C) and be fully open at 207° F (97° C).

WATER PUMP

All Models

71. R&R AND OVERHAUL. To remove the water pump, first remove hood and front side panels. Drain the cooling system and disconnect upper and lower radiator hoses. Loosen adjusting bolts on alternator and air conditioning compressor, if so equipped, and remove belts. Unbolt fan shroud and move shroud rearward, then unbolt and remove fan and pulley. Remove mounting bolts and lift out water pump.

Disassemble the pump as follows: Remove plastic screw (13—Fig. 61), then using a 1/2 × 2 inch NC cap screw for a jack screw, force impeller (12) off rear end of shaft. Using two screwdrivers, pry seal assembly (10) from pump body. Support hub (4) and press out shaft. Press shaft and bearing assembly (5) out front of body. Make certain that body is supported as close to bearing as possible.

When reassembling, press shaft and bearing assembly into body using a piece of pipe so pressure is applied only to outer race of bearing. Bearing race should be flush with front end of body. Install new seal assembly (10). Press only on outer diameter of seal. Clean inside of hub (4) with degreasing solvent. Apply a light coat of Loctite to both surfaces. Support shaft assembly and press hub on shaft until hub is flush with end of shaft. Install face ring (11) and ''O'' ring (9) in impeller (12), then press impeller on shaft until there is a clearance of 0.012-0.020 inch (0.305-0.508 mm) between body and front of impeller. See Fig. 62. Install plastic screw (13—Fig. 61).

Using a new gasket (8) reinstall pump by reversing removal procedure. Tighten water pump retaining bolts to a torque of 15 ft.-lbs. (20 N·m).

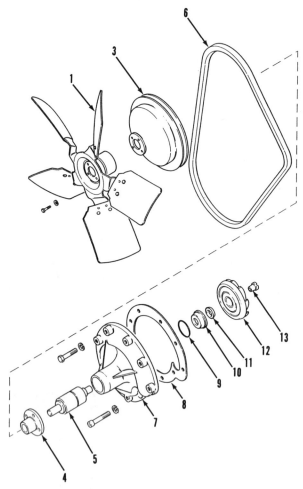

Fig. 61—Exploded view of typical water pump assembly. The five blade fan is standard. A six blade fan is used on models equipped with air conditioning.

1. Fan	
3. Pulley	8. Gasket
4. Hub	9. ''O'' ring
5. Shaft & bearing	10. Seal assy.
6. Water pump-	11. Face ring
alternator belt	12. Impeller
7. Pump body	13. Plastic screw

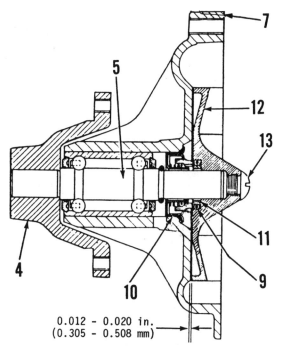

0.012 - 0.020 in.
(0.305 - 0.508 mm)

Fig. 62—Sectional view of typical water pump, showing correct dimension for installation of impeller.

ELECTRICAL SYSTEM

BATTERIES

All Models

72. Before any electrical service is performed, a thorough check of battery condition, condition of cable connections and condition of alternator belt and pulleys should be made.

Batteries should be checked for total voltage and voltage drop under rated load. All relative connections should be checked for excessive resistance using an ohmmeter.

If voltage of batteries is excessively low, they should be recharged to rated level, using an external battery charger. Failure to do so can cause alternator to overheat resulting in premature failure of alternator, regulator or both.

73. BATTERY CURRENT DRAIN TEST. If after checking, servicing and installing batteries, they return to an undercharged condition when tractor is not in use, current drain should be suspected. To check, disconnect negative battery terminal. Connect an ammeter lead to negative battery post and the other lead to negative cable end. Make certain all electrical systems are off. Any reading on ammeter indicates current leakage. Check for lights, radio or other accessories being on, or for shorted electrical wiring or components.

CHARGING SYSTEM

All Models

74. TESTING. Prior to beginning tests, be sure batteries are fully charged, all connections are clean and tight and alternator drive belt is properly adjusted.

CAUTION: Do not disconnect any wires from alternator while engine is running as damage to alternator may result.

Alternator warning light should not come on when key switch is in off position. If light fails to go out when key is in off position, test circuits as outlined in paragraph 75.

Normal operation of the warning light is to come on only when key switch is in first clockwise position from off, engine not running. If engine is started while throttle lever is in low idle position, light may stay on until engine speed is increased for the first time. Light should not come on at any time after engine is running and initial engine speed increase was sufficient to shut light off. If operation of warning light is not normal, test circuits as outlined in paragraph 75.

75. WARNING LIGHT CIRCUIT TEST. If warning light fails to go out with key switch in off position, disconnect plug at alternator. If light goes out after plug is removed from alternator, renew rectifier assembly in alternator.

If light stays on after disconnecting plug at alternator, there is an electrical short between alternator wires and another positive wire in wiring harness. Repair as necessary.

If light fails to come on with key switch in first position clockwise from off, engine not running, disconnect plug from alternator and connect a jumper wire to thin wire in connector and ground other end of jumper wire to alternator housing. If light comes on at this time, alternator light bulb and wiring are good and problem is in the alternator. Check as outlined in paragraph 76. If light still fails to come on, check bulb, bulb socket, current supply to socket and wiring from socket to alternator.

If warning light functions normally with engine not running, but fails to go off when engine is running at various speeds, check alternator as outlined in paragraph 76.

76. ALTERNATOR VOLTAGE OUTPUT TEST. Make certain all wiring connections are in place. Then connect voltmeter positive lead to BAT terminal of alternator and ground negative lead to adequate ground. Start engine and operate at 1800 rpm. Voltmeter should read between 13 and 15 volts. If voltage is less than 13 volts, bypass regulator and check as outlined in paragraph 76A.

Voltage in excess of 15 volts indicates either a grounded brush inside alternator or faulty voltage regulator. Repair as necessary.

76A. VOLTAGE REGULATOR BYPASS TEST. With voltmeter connected as outlined in paragraph 76, start and operate engine at 1800 rpm. Ground voltage regulator to alternator housing using a probe with ground wire inserted through test hole in back of alternator cover as shown in Fig. 63. If voltage before bypassing voltage regulator was not between 13 and 15 volts, but is now between 13 and 15 volts with regulator bypassed, renew voltage regulator. If voltage fails to be between 13 and 15 volts while bypassing regulator, remove and overhaul alternator.

77. ALTERNATOR AMPERAGE TEST. Connect an ammeter between battery terminal on starter motor and the output wire (brown/white wire on models without cab or red wire on models with cab) at rear of alternator. Connect a variable resistance across battery. Apply a load to the battery which is just above the maximum alternator rated output. Start engine and slowly increase engine speed to high idle rpm. (Refer to paragraph 62.) The reading on the ammeter must be the maximum rated output. If the ammeter reading is not correct, the fault is in the surge protection diode, rotor winding, stator winding, rectifier or brushes.

ALTERNATOR

78. Lucas alternators are used on all models. Lucas Models A115-45 and A127-45 are used on models not equipped with cab and Model A127-65 is used on models with cab. All alternators are equipped with

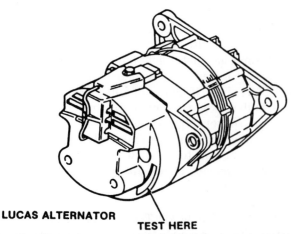

LUCAS ALTERNATOR

TEST HERE

Fig. 63—View showing test hole for bypassing voltage regulator on typical Lucas alternator.

nonadjustable solid state type voltage regulators. Voltage regulator is internally mounted on Model A115-45 alternator and externally mounted on Model A127-45 and A127-65. Specifications for alternators is as follows:

Lucas A115-45

Alternator type	12V negative ground
Maximum output at 6000 rpm	45.0 amperes
Stator type	3-phase Star connected
Winding resistance (each phase)	3.2 ohms
New brush length	0.8 in. (20.0 mm)
Minimum brush length	0.4 in. (10.0 mm)
Brush spring pressure	4.7-9.8 oz. (1.3-2.7 N)

Lucas A127-45

Alternator type	12V negative ground
Maximum output at 6000 rpm	45.0 amperes
Stator type	3-phase Delta connected
Winding resistance (each phase)	3.2 ohms
New brush length	0.64 in. (16.3 mm)
Minimum brush length	0.2 in. (5.0 mm)
Brush spring pressure	4.7-9.8 oz. (1.3-2.7 N)

Lucas A127-65

Alternator type	12V negative ground
Maximum output at 6000 rpm	65.0 amperes
Stator type	3-phase Delta connected
Winding resistance (each phase)	3.2 ohms
New brush length	0.64 in. (16.3 mm)
Minimum brush length	0.2 in. (5.0 mm)
Brush spring pressure	4.7-9.8 oz. (1.3-2.7 N)

CAUTION: Because certain components of the alternator can be damaged by procedures that will not affect a DC generator, the following precautions MUST be observed.
 a. **When installing batteries or connecting a booster battery, negative post of batteries must be grounded.**
 b. **Never short across any terminal of the alternator.**
 c. **Do not attempt to polarize the alternator.**
 d. **Disconnect all battery ground straps before removing or installing any electrical unit.**
 e. **Do not operate alternator on an open circuit and make certain all leads are properly connected before starting engine.**

Lucas A115-45

79. DISASSEMBLY. To disassemble the alternator, unbolt and remove end cover (15—Fig. 64). Disconnect yellow and black wires, then remove regulator (19). Remove outer brush screw and remove brushes (20). Unbolt and remove brush box (17). Place an alignment chalk mark across drive end frame (5), stator (22) and slip ring end frame (21). Remove the three through-bolts and remove stator (22) and slip ring end frame (21) assembly. Unsolder stator wires from rectifier (16), remove three mounting screws and lift off rectifier (16). Remove nut (1), lockwasher (2), pulley (3) and fan (4), then press rotor (12) from drive end frame. Remove snap ring (11) and dished washer (10), then press bearing (9), seal ring (8),

dished washer (7) and spacer (6), from drive end frame (5).

80. TESTING. To test the rotor windings, connect an ohmmeter between slip ring connectors as shown in Fig. 65. Ohmmeter should read 3.2 ohms. If not, renew rotor. To check insulation of rotor windings, connect a 15 watt test light in a 110 volt circuit connected to outer slip ring and any rotor pole as shown in Fig. 66. If light comes on, renew rotor.

To renew rear bearing (13—Fig. 64), unsolder slip ring connecting wires and remove slip rings (14). Press rear bearing off the rotor.

Continuity of stator windings can be checked by connecting any two of the three stator leads in series with a 12 volt test light of not less than 36 watts. If test light comes on, transfer one of the test light leads to remaining stator winding lead. See Fig. 67. Test light should come on at each position. If not, renew stator assembly.

Insulation of stator winding may be checked by connecting a 110 volt, 15 watt, AC current test light lead to stator laminations and touch each of the stator leads in turn with remaining test light lead. See Fig. 68. If test light comes on, stator is short circuited and must be renewed.

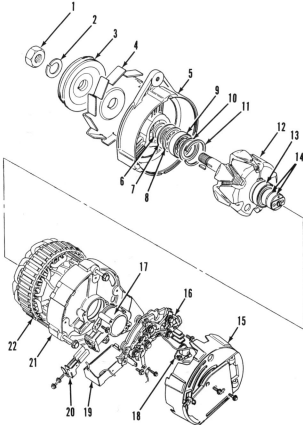

Fig. 64—Exploded view of Lucas Model A115-45 alternator.

1. Nut		13. Rear bearing
2. Lockwasher		14. Slip rings
3. Pulley		15. Rear cover
4. Fan		16. Rectifier
5. Drive end frame		17. Brush box
6. Spacer		18. Surge protection
7. Dished washer		diode
8. Seal ring		19. Regulator
9. Bearing		20. Brushes
10. Dished washer		21. Slip ring end
11. Snap ring		frame
12. Rotor		22. Stator

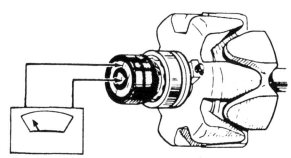

Fig. 65—Use an ohmmeter between slip ring connectors to check continuity of rotor windings on Model A115-45 alternator.

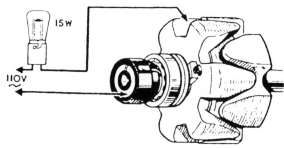

Fig. 66—To check insulation of rotor windings, connect a 15 watt light bulb in a 110 volt circuit connected to outer slip ring and any rotor pole.

Diodes in rectifier act as one-way gates for electrical current by allowing current to pass through in one direction only. They may be checked using a 12 volt battery and a 1.5 watt (maximum) test light connected between diode and heat sink. See Fig. 69. Observe light. Reverse lead connections and observe light. Light should be on in one direction only. Test each of the six diodes with its respective heat sink. If any diode tests faulty, renew rectifier assembly.

Check length of brushes. New brushes are 0.8 inch (20.0 mm) long. Renew brushes if worn to a length of less than 0.4 inch (10.0 mm). Install brushes in brush box and using special tool as shown in Fig. 70, check brush springs. It should require a pressure of 4.7-9.8 ounces (1.3-2.7 N) to push each brush inward.

81. REASSEMBLY. Reassemble by reversing the disassembly procedure, keeping the following points

in mind: Note that some type of heat sink should be used between diodes and connections to be resoldered (needlenose pliers lightly clamped on diode lead) to protect diode from heat damage. Use the following torques when reassembling:

Pulley nut	20-35 ft.-lbs. (27.2-27.5 N·m)
Through-bolts	3.3-4.6 ft.-lbs. (4.5-6.2 N·m)
Brush box bolts	2.5-2.9 ft.-lbs. (3.4-4.0 N·m)
Outer brush screw	1.3-1.7 ft.-lbs. (1.7-2.3 N·m)
Regulator screws	1.3-1.7 ft.-lbs. (1.7-2.3 N·m)
Rectifier bolts	2.5-2.9 ft.-lbs. (3.4-4.0 N·m)
End cover bolts	1.3-1.7 ft.-lbs. (1.7-2.3 N·m)

Lucas A127-45 And A127-65

82. DISASSEMBLY. To disassemble either alternator, first unbolt and remove the regulator and

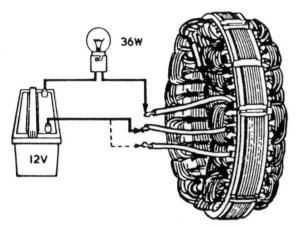

Fig. 67—Continuity check of stator windings using a 36 watt test light and a 12 volt circuit.

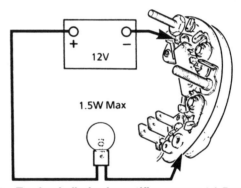

Fig. 69—To check diodes in rectifier, connect 1.5 watt test light as shown.

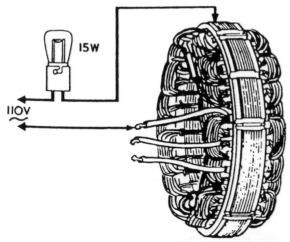

Fig. 68—Use a 15 watt test light and 110 volt circuit to check condition of stator winding insulation.

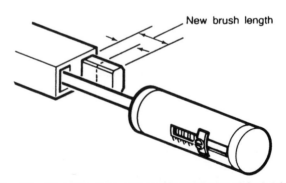

Fig. 70—New brush length on Model A115-45 is 0.8 inch (20.0 mm). Renew brushes if length is less than 0.4 inch (10.0 mm). Use special tool as shown to check brush springs.

brush holder assembly (1—Fig. 71). Place an alignment chalk mark across slip ring end frame (2), stator frame (10) and drive end frame (9). Remove the three through-bolt nuts and lockwashers and separate slip ring end frame (2) and stator (10) from drive end frame and rotor. Remove nuts, washers and insulators from stud terminals and the two screws securing stator to rear frame, then remove stator. Unsolder and remove stator leads from rectifier and separate rectifier from stator.

Remove pulley nut (5), lockwasher (6), pulley (7), fan (8), spacer (16) and washer (15) from rotor shaft. Remove rotor (11) and spacer (13) from drive end frame (9) and bearing (14) assembly. If bearing (14) is faulty, renew drive end frame assembly as bearing is not serviced separately. If rear needle bearing (3) is excessively worn, press out old bearing and press in new bearing until flush with frame.

83. TESTING. To check the rotor windings for continuity connect an ohmmeter between slip rings

as shown in Fig. 72. Ohmmeter reading should be 3.2 ohms. If not, renew rotor. To check insulation of rotor windings, connect a 15 watt test light in a 110 volt circuit connected to outer slip ring and any rotor pole as shown in Fig. 73. If light comes on, renew rotor.

Continuity of stator windings can be checked by connecting any two of the three stator leads in series with a 12 volt test light of not less than 36 volts. See Fig. 67. If test light comes on, transfer one of the test light leads to the remaining stator winding lead. Test light should come on at each position. If not, renew stator assembly.

Insulation of stator windings can be checked by connecting a 110 volt, 15 watt, AC current test light lead to stator laminations and touch each of the stator leads in turn with remaining test light lead. See Fig. 68. If test light comes on, stator is short circuited and must be renewed.

Diodes in rectifier act as one-way gates for electrical current by allowing current to pass through in one direction only. They can be checked using a 12 volt battery and a 1.5 watt (maximum) test light connected between diode and heat sink. See Fig. 69. Ob-

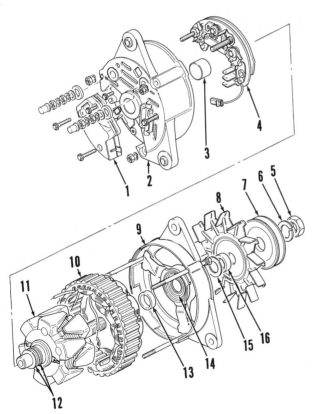

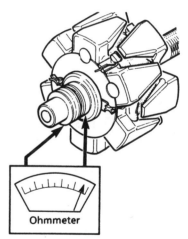

Fig. 72—Connect an ohmmeter between slip rings to check continuity of rotor windings on Model A127-45 or A127-65 alternators.

Fig. 71—Exploded view of Lucas Model A127-45 or A127-65 alternators.

1. Regulator & brush holder assy.
2. Slip ring end frame
3. Rear bearing
4. Rectifier
5. Nut
6. Lockwasher
7. Pulley
8. Fan
9. Drive end frame
10. Stator
11. Rotor
12. Slip rings
13. Spacer
14. Bearing
15. Washer
16. Spacer

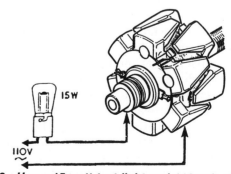

Fig. 73—Use a 15 watt test light and 110 volt circuit to check condition of stator winding insulation.

serve light. Reverse lead connections and observe light. Light should be on in one direction only. Test each of the six diodes with its respective heat sink. If any diode tests faulty, renew rectifier assembly.

Check length of brushes. New brushes are 0.64 inch (16.3 mm) long. Renew brushes if worn to a length of less than 0.2 inch (5.0 mm). Install brushes in brush holder and using special tool as shown in Fig. 70, check brush springs. It should require a pressure of 4.7-9.8 ounces (1.3-2.7 N) to push each brush inward.

84. REASSEMBLY. Reassemble by reversing the disassembly procedure, keeping the following points in mind: Note that some type of heat sink should be used between diodes and connections to be resoldered (needlenose pliers lightly clamped on diode lead) to protect diode from heat damage. Use the following torques when reassembling:

Pulley nut	27.5 ft.-lbs.
	(37.5 N·m)
Through-bolt nuts	4.0 ft.-lbs.
	(5.5 N·m)
Rear frame to stator screws	2.5 ft.-lbs.
	(3.5 N·m)
Rectifier terminal nuts	3.0 ft.-lbs.
	(4.0 N·m)
Regulator & brush holder screws	2.0 ft.-lbs.
	(2.5 N·m)

STARTING MOTOR

Lucas or Bosch starting motors have been used on all models. Refer to appropriate following paragraphs for service information.

Lucas Starter

85. DISASSEMBLY AND TESTING. To disassemble the Lucas starter, first remove solenoid assembly and copper link connecting solenoid to starter. Remove through-bolts (30—Fig. 74) and end cover (29). Use caution to avoid losing brake shoes (25) and springs (24). Remove brushes (23) from brush holder (22) and remove the holder. Inspect brush springs (21) and renew as needed. Remove field housing (18). Remove eccentric pin (1) and remove drive end housing (3). Push thrust collar (12) off the retaining ring (11), then remove the ring and collar. Remove starter drive (13), center bearing plate (14) and shims (16). Note thickness and number of shims (16) for aid in reassembly. Inspect bushings (2, 15 and 27) for excessive wear or other damage. Renew as required. Soak new bushings in oil for 24 hours prior to installation.

86. ARMATURE. Inspect armature commutator and if worn, rough or pitted, it can be trued in a lathe. Minimum commutator diameter is 1.50 inches (38.1 mm). Polish commutator with nonconductive emery cloth. Do not undercut insulators between commutator segments.

87. ARMATURE INSULATION TEST. Armature insulation can be tested by using a 110 volt, 15 watt, AC current test light connected as shown in Fig. 75. Touch each commutator segment in turn with test light lead. If test light comes on at any segment, windings are shorted and must be repaired or renewed.

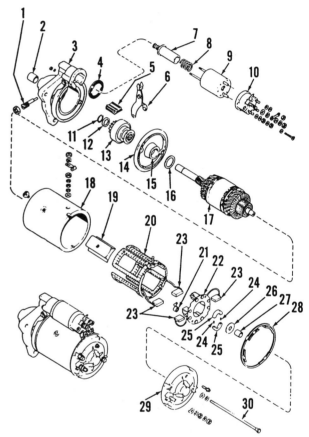

Fig. 74—Exploded view of typical Lucas starter motor assembly.

1. Pivot bolt	15. Bushing
2. Bushing	16. Shims
3. Drive end housing	17. Armature
4. Gasket	18. Field housing
5. Grommet	19. Pole shoe
6. Shift link	20. Field coils
7. Plunger	21. Brush springs
8. Spring	22. Brush holder
9. Solenoid coil	23. Brushes
10. End cap	24. Brake springs
11. Retaining ring	25. Brake shoes
12. Collar	26. Thrust washer
13. Starter drive	27. Bushing
14. Center bearing plate	28. Seal
	29. End cover
	30. Through-bolt

88. FIELD COIL TEST. Field coil insulation can be checked using 110 volt, 15 watt, AC current test light connected as shown in Fig. 76. Brushes must not touch field housing and power terminal insulation must not allow voltage to short to housing. If test light comes on and brushes are clear of housing and power terminal insulation is good, there is a short between field coils (20—Fig. 74) and field housing (18). Repair or renew as necessary.

89. INSULATED BRUSH BOX TEST. Two of the brush boxes on brush holder (22—Fig. 74) must be insulated from the assembly itself. Connect a 110 volt, 15 watt, AC current test light as shown in Fig. 77 and touch test light lead to each insulated brush box in turn. If test light comes on, insulation between brush box and brush holder assembly is faulty and brush holder assembly must be renewed.

90. REASSEMBLY. Reassemble by reversing disassembly procedure. Make certain slots in brake shoes (25—Fig. 74) align with cross pin in armature shaft. Armature end play must be 0.005-0.025 inch (0.13-0.63 mm) and is controlled by thickness of shims (16) between center bearing plate (14) and armature core.

Pinion clearance is set by connecting a 6 volt battery to solenoid on assembled starter to activate sole-

noid. Push back lightly on starter drive clutch (1—Fig. 78) to remove any free play. Measure with a feeler gage as shown in Fig. 78. Turn eccentric pivot bolt (1—Fig. 74) to adjust clearance and when set, lock pivot bolt in place with locknut. Pinion clearance should be 0.005-0.045 inch (0.13-1.14 mm).

Bosch Starter

91. DISASSEMBLY AND TESTING. To disassemble Bosch starter, first remove solenoid assembly (1—Fig. 79) and shift link pivot pin. Remove end cap (16), seal ring (15), "U" retainer (14) and shims (13). Remove through-bolts (18) and end cover (12). Lift brushes (9) from brush boxes and remove brush holder (10). Remove field housing (7) and drive hous-

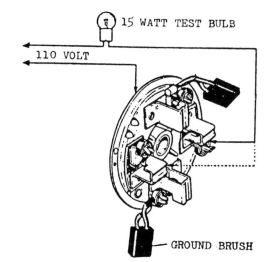

Fig. 77—Check both insulated brush holders as shown.

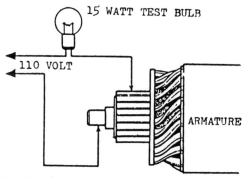

Fig. 75—When checking armature winding insulation, each commutator segment must be checked separately.

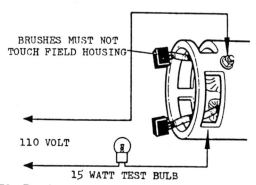

Fig. 76—Brushes must not touch housing and insulation at terminal must be good when checking field coils.

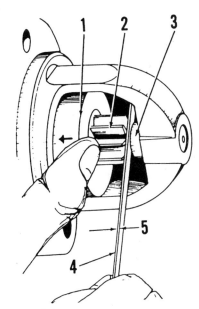

Fig. 78—Measure pinion clearance with a feeler gage as shown.

ing (3). Remove starter drive (23) and center bearing plate (5). Inspect bushings (2, 4, 11 and 22) for excessive wear or other damage. Renew as required. Soak new bushings in oil for 24 hours prior to installation. Inspect brushes and springs and renew if excessively worn or damaged. Brush minimum length is 0.335 inch (8.5 mm). The new service brushes are attached with small bolts instead of solder. Replacement brush service package (19) consists of four brushes and attaching bolts.

92. ARMATURE. Inspect armature commutator for wear, roughness or pitting. Commutator can be trued in a lathe. Minimum commutator diameter is 1.67 inches (42.5 mm). The mica insulation between segments can be undercut to a depth of 0.001 inch (0.025 mm). After undercutting, polish commutator with nonconductive emery cloth.

93. ARMATURE INSULATION TEST. Armature insulation can be checked by using a 110 volt, 15 watt test light connected as shown in Fig. 75. Touch each commutator segment in turn with test light lead. If test light comes on at any segment, windings are shorted and armature must be repaired or renewed.

94. FIELD COIL TEST. Inspect field coils for overheating and burnt insulating wrap, then test as follows: Connect a 110 volt, 15 watt test light as shown in Fig. 76. Brushes must not touch field housing and power terminal insulation must not allow voltage to short to housing. If test light comes on and brushes are clear of housing and power terminal insulation is good, there is a short between field coils (6—Fig. 79) and field housing (7). Repair or renew as necessary.

95. INSULATED BRUSH BOX TEST. Two of the brush boxes on brush holder (10—Fig. 79) must be insulated from the assembly itself. Connect a 110 volt, 15 watt test light as shown in Fig. 77 and touch test light lead to each insulated brush box in turn. If test light comes on, insulation between brush box and brush holder is faulty and brush holder assembly must be renewed.

96. REASSEMBLY. Reassemble by reversing the disassembly procedure. Armature end play is controlled by shims (13—Fig. 79). Before installing end cap (16), use a dial indicator and check armature end play. Add or remove shims (13) as required to obtain end play of 0.004-0.016 inch (0.1-0.4 mm). Install "U" retainer (14), new seal ring (15) and end cap (16).

NOTE: Use Bosch caulking wax spray FL58V3 or equivalent to waterproof all housing joints during reassembly.

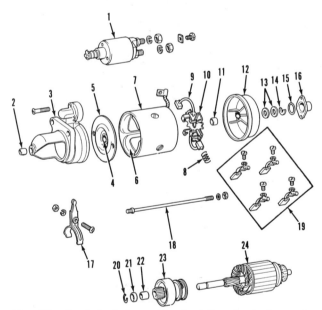

Fig. 79—Exploded view of typical Bosch starter motor assembly.

1. Solenoid assy.	13. Shims
2. Bushing	14. "U" retainer
3. Drive housing	15. Seal ring
4. Bushing	16. End cap
5. Center bearing plate	17. Shift link
6. Field coils	18. Through-bolt
7. Field housing	19. Brush service package
8. Brush springs	20. Retaining ring
9. Brushes	21. Collar
10. Brush holder	22. Bushing
11. Bushing	23. Starter drive
12. End cover	24. Armature

CLUTCH

Models 385, 485 and 585 are equipped with a Laycock pressure plate assembly and an 11 inch (279 mm) single plate dry disc engine clutch plate.

Models 685 and 885 are equipped with a Rockford pressure plate assembly and a 12 inch (305 mm) single plate dry disc engine clutch plate.

On models not equipped with cab, all tractors have mechanical clutch operating linkage. On models equipped with cab, engine clutch is hydraulically operated.

ADJUSTMENT

Models Without Cab

97. PEDAL HEIGHT. Clutch linkage is identical on all models without a cab. However, adjustment dimensions for Laycock 11 inch (279 mm) clutch differs from dimensions for Rockford 12 inch (305 mm) clutch.

To adjust the clutch linkage, loosen locknut (2—Fig. 80). Remove return spring and clevis pin. Adjust clevis (3) on rod until the following dimensions are obtained with clevis connected:

Laycock Clutch
C. Free travel . . . . .1-1/4 in.
(32 mm)
E. Full travel . . . . .4-17/32 in.
(115 mm)
D. . . . . .6-5/16 in.
(160 mm)

Rockford Clutch
C. Free travel . . . . .1/2 in.
(12.7 mm)
E. Full travel . . . . .4-21/32 in.
(118 mm)
D. . . . . .7-1/16 in.
(179 mm)

Reconnect return spring and tighten locknut. Loosen locknut and adjust safety start switch (5) until there is a clearance between safety start switch and clutch pedal of 9/32 inch (7 mm). Tighten locknut.

Models With Cab

98. PEDAL HEIGHT. To adjust the clutch pedal height on models with cab, disconnect line from master cylinder. Unbolt master cylinder and loosen locknut (1—Fig. 81) on push rod (4). Loosen locknut on stop bolt (2), then turn stop bolt counterclockwise several turns. Raise master cylinder and rotate push rod with fingers as required. Install master cylinder and tighten retaining bolts. Measure distance from floor pan (mat removed) to pedal as shown in Fig. 81.

Distance should be 7-7/8 to 8-1/8 inch (199 to 205 mm). If not, readjust push rod to obtain correct dimension. Then, tighten locknut on push rod. Turn stop bolt (2) until it contacts clutch pedal, turn bolt clockwise an additional 1/4 turn and tighten locknut.

Check safety start switch position. Loosen locknut and turn switch in or out as required so that switch button is pushed inward when clutch pedal is fully depressed. Tighten locknut. Reconnect lines to master cylinder. Fill reservoir and bleed air from system as outlined in paragraph 100.

99. CLUTCH CYLINDERS. To remove clutch master cylinder, first disconnect battery ground cable. Then, press inward on lock pins and remove instrument panel front cover. Remove screws from both sides of instrument panel, raise and support instrument panel. Disconnect pedal return spring. Place a container under instrument panel to catch the fluid. Disconnect lines from master cylinder. Unbolt master cylinder from bracket and raise the cylinder. Loosen locknut and turn push rod counterclockwise from the clevis. Lift out master cylinder.

To disassemble the master cylinder, first remove locknut from push rod (7—Fig. 82). Remove rubber boot (10), snap ring (9), retaining washer (8) and push rod (7). Remove plunger and spring assembly from cylinder housing (1). Remove seals (2 and 6) from plunger assembly. Clean and inspect all parts. A service kit consisting of seal (1), wave washer (4), seal (6), retaining washer (8), snap ring (9) and rubber boot (10) is available. If plunger or cylinder housing shows

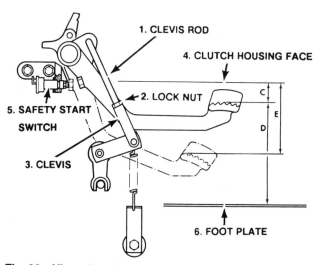

Fig. 80—View of engine clutch linkage for models without cab. Adjustment dimensions for Laycock 11 inch (279 mm) clutch differs from dimensions for Rockford 12 inch (305 mm) clutch.

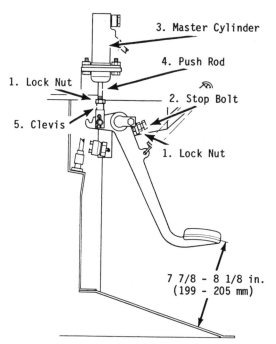

Fig. 81—View showing engine clutch pedal height adjustment on models equipped with cab. Refer to text for procedure.

signs of scoring or other damage, renew cylinder assembly. Reassemble and reinstall by reversing disassembly and removal procedures. Bleed air from system as in paragraph 100. Adjust pedal height as in paragraph 98.

To remove clutch slave cylinder, remove floor mat and inspection panel from floor. Disconnect hose from slave cylinder. Remove linkage return spring. Remove the two snap rings and linkage pin. Unbolt and remove slave cylinder.

To disassemble the slave cylinder, remove eye clevis (7—Fig. 83). Remove rubber boot (6) and snap ring (5), then withdraw piston (4) with seal (3). Remove seal from piston. Clean and inspect piston and cylinder housing (2) for scoring or other damage. If scored or damaged, renew cylinder assembly. A service kit consisting of bleed screw rubber cap (1), seal (3), snap ring (5) and rubber boot (6) is available. Reassemble and reinstall by reversing the disassembly and removal procedures. When installing eye clevis (7) on piston (4), apply Loctite 270 to first three threads of piston rod. Adjust eye clevis so that a distance of 3.4 inches (86.0 mm) is obtained between centers of cylinder mounting holes and eye clevis hole (with piston fully inward). Bleed air from system as outlined in paragraph 100.

100. BLEED CLUTCH SYSTEM. To bleed air from system, remove bleed screw cap (1—Fig. 83) and connect a length of transparent hose with its free end in Hy-Tran Plus fluid. Loosen bleed screw. Fill reservoir with Hy-Tran Plus MS 1207 fluid to maximum level mark. Fully press and release clutch pedal. Repeat as required until fluid free of air bubbles flows through the hose. Then, tighten bleed screw while holding pedal down.

NOTE: Make certain that fluid level in reservoir is kept above minimum level mark during bleeding operation.

Remove bleed hose and refill reservoir with fluid.

R&R AND OVERHAUL CLUTCH

All Models

101. To remove engine clutch, it is first necessary to detach (split) engine from clutch housing as follows: Disconnect battery cables and remove battery. Remove hood, grille and side panels. Disconnect tachometer cable, starter wiring and rear wiring connector. Shut off fuel and disconnect supply line and fuel return line. Cap all openings. Disconnect injection pump speed control and shut off cables. Identify and disconnect oil cooler lines and steering lines and cap all openings. On models so equipped, disconnect heater hoses and air conditioning lines. On models equipped with front drive axle, remove drive shaft shield and front drive shaft. If available, attach a split stand under clutch housing and engine oil pan or block up under clutch housing and attach a hoist to engine. Place wood blocks between front axle and

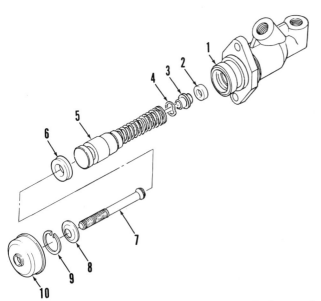

Fig. 82—Exploded view of clutch master cylinder used on models equipped with cab.

1. Housing	6. Seal
2. Seal	7. Push rod
3. Valve stem	8. Retaining washer
4. Wave washer	9. Snap ring
5. Plunger	10. Rubber boot

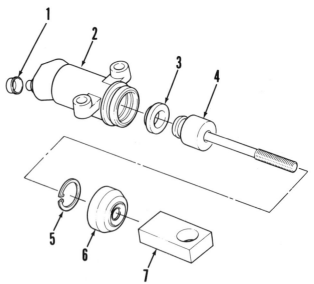

Fig. 83—Exploded view of clutch slave cylinder used on models equipped with cab.

1. Bleed screw cap		
2. Housing		5. Snap ring
3. Seal		6. Rubber boot
4. Piston		7. Eye clevis

front support to prevent tipping. Unbolt clutch housing from engine and move front half of tractor forward.

Unbolt and remove pressure plate assembly and remove clutch friction disc. Inspect the friction disc and if pads are worn to a point where there is less than 0.010 inch (0.24 mm) of material above rivet heads, renew the disc. Inspect pressure plate for grooving

and flatness. If plate is grooved more than 0.025 inch (0.64 mm) or if face is worn out-of-flat more than 0.006 inch (0.152 mm), renew pressure plate. Do not reface a worn pressure plate.

When reinstalling, reverse removal procedure. Clutch friction disc is marked FLYWHEEL SIDE for correct installation. Adjust clutch pedal height as outlined in paragraph 97 or 98.

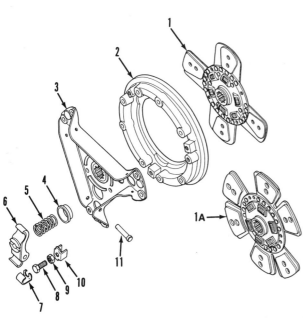

Fig. 84—Exploded view of 11 inch (279 mm) Laycock clutch. Four pad clutch disc is used on Model 385. Six pad clutch disc is used on Models 485 and 585.

1. Clutch disc (385)	
1A. Clutch disc (485 & 585)	6. Finger
	7. Spring clip
2. Pressure plate	8. Cap screw
3. Cover	9. Nut
4. Spring cup	10. Wear clip
5. Spring	11. Pin

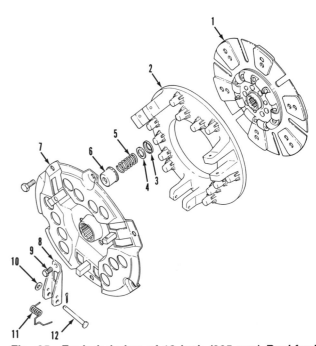

Fig. 85—Exploded view of 12 inch (305 mm) Rockford clutch used on Models 685 and 885.

1. Clutch disc	
2. Pressure plate	8. Finger
3. Spring seat	9. Cap screw
4. Insulating washer	10. Washer
5. Spring	11. Finger tension spring
6. Spring cup	
7. Cover	12. Pin

TWO SPEED POWER SHIFT

OPERATION

Models So Equipped

102. All models may be equipped with a two speed Power Shift unit located in front of the speed transmission. The two speed unit is controlled by an electric solenoid/hydraulic valve. The two position control switch is located on left side of instrument panel. When shifted from high (direct) drive to low drive, tractor power is increased with a 17 percent reduction in speed in any tractor gear without interruption of engine power. This gives the tractor a total of sixteen forward speeds and eight reverse speeds.

REGULATOR (CONTROL) VALVE

Models So Equipped

103. R&R AND OVERHAUL. The regulator (control) valve (4—Fig. 86) and bottom core plate (12) can be removed from tractor without removing the speed gearbox. All other service on the Power Shift unit requires removal of complete speed gearbox.

To remove the regulator valve and bottom core plate, first disconnect battery ground cable. Remove drain plug and drain transmission fluid. On models equipped with front drive axle, unbolt and remove drive shaft shield and front drive shaft. On all models, unbolt and remove solenoid wiring shield (11—Fig. 86). Disconnect the Power Shift cable (8) from solenoid cable (9) at connector. Unbolt bottom core plate (12) and remove from speed gearbox.

Disconnect wiring from solenoid, then unscrew union and remove wiring and union from core plate (12). Remove screws (4—Fig. 87), clamp (5) and solenoid (1) with "O" rings (2 and 3). Remove regulator valve end cap (14), washer (13), pin (12), spring (11) and valve (10). Remove plug (9), spring (8) and spool (7). Unbolt and remove regulator valve body (6), then remove filter screen (1—Fig. 86) from bottom of valve body. Remove gasket (15), separator plate (14) and

gasket (13) from core plate (12). Remove oil inlet tube (7). Oil tubes (6) will remain in speed gearbox housing.

Clean and inspect all parts for excessive wear or other damage and renew as necessary. Body (6—Fig. 87), spool (7) and plug (9) are not serviced separately. When reassembling, use new "O" rings (2 and 3) and new gaskets (13 and 15—Fig. 86) and "O" rings (5) and reverse disassembly procedures.

POWER SHIFT UNIT

Models So Equipped

104. R&R AND OVERHAUL. To remove the Power Shift unit, speed gearbox must be removed from the tractor. To remove the speed gearbox, first remove cab and fuel tanks or fenders, tank and platform, as equipped, as outlined in paragraph 145 or 144. Remove plugs and drain fluid from speed gearbox and rear frame. Remove front drive axle drive shaft and shield, if so equipped. Unbolt and remove pto clutch cover from bottom of rear main frame. Refer to Fig. 88 and drive out roll pin (1) from pto clutch and shaft. Unbolt lower pto shaft cover or seal housing, as equipped, and remove with gasket from rear of tractor. Have helper support the pto clutch. With-

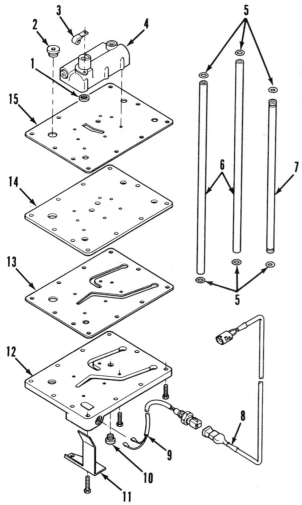

Fig. 86—View of Power Shift regulator (control) valve and bottom core plate with relative components removed from speed transmission gearbox.

1. Screen	8. Power shift cable
2. Plug	9. Solenoid cable
3. Wire clip	10. Plug
4. Regulator (control) valve assy.	11. Solenoid wire shield
5. "O" rings	12. Bottom core plate
6. Pressure tubes	13. Gasket
7. Oil inlet tube	14. Separator plate
	15. Gasket

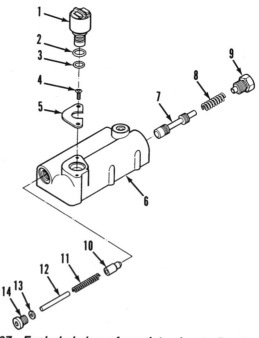

Fig. 87—Exploded view of regulator (control) valve and solenoid.

1. Solenoid	8. Spring
2. "O" ring	9. Plug
3. "O" ring	10. Valve
4. Screw (2)	11. Spring
5. Clamp	12. Pin
6. Valve body	13. Washer
7. Spool	14. Plug

draw lower pto shaft and remove pto clutch. Remove the snap ring from rear of pto driven gear, then remove driven gear. Disconnect 1st-2nd and 3rd-4th selector rods from gear selector levers.

Disconnect all interfering hydraulic lines and wiring. Using a sling and hoist (Fig. 89), lift off steering support assembly. Remove right and left brake pipes. Remove steering supply and return pipes, power shift return pipe and oil cooler supply and return pipes.

NOTE: Plug or cap all openings immediately to prevent foreign material from entering systems.

Place special tool (Churchill #MS 2700) or equivalent, under tractor. See Fig. 89A. Adjust splitting tool to support tractor under speed gearbox and rear main frame. Place wooden wedges between front axle and front support to prevent tipping. Unbolt range section (rear frame) from speed gearbox. After making certain that all wires and hydraulic lines are disconnected, roll engine and speed gearbox forward from range section.

Attach a sling and hoist to speed gearbox and block up securely under engine oil pan. Unbolt clutch housing from engine, then separate speed gearbox from engine. If available, attach speed gearbox assembly in a transmission stand.

Unbolt and remove bottom core plate (12—Fig. 86) and regulator valve (4) assembly, then remove oil inlet tube (7). Remove clutch release bearing and sleeve. Drive roll pins from release fork, withdraw release shaft and remove fork. Unbolt and remove pto driven shaft end cover and metal gasket as shown in Fig. 90. Remove snap ring from front end of driven shaft. Working through bottom core plate opening, remove snap ring from pto driven gear. Drive the driven shaft rearward out of bearing and remove bearing. Withdraw the shaft and have helper remove gear as shaft is removed. Unbolt and remove clutch release sleeve carrier, then withdraw pto drive shaft and gear as shown in Fig. 91. Remove snap ring from front end of countershaft and remove countershaft gear. Remove the two speed Power Shift unit from front of speed gearbox. See Fig. 92.

If removal of direct drive gear (2—Fig. 95) is desired, refer to paragraph 110 and remove speed selector cover, selector shaft and forks and the mainshaft and gears. Then, see Figs. 93 and 94.

Fig. 88—Bottom view of pto clutch (cover removed).

1. ROLL PIN
2. SHAFT
3. P.T.O. CLUTCH ASSEMBLY

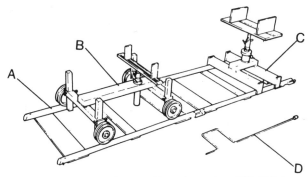

Fig. 89A—View of Churchill tool number MS 2700 used for splitting tractor.

A. Guide rails
B. Movable support
C. Static support
D. Adjusting crank

Fig. 89—Attach sling and hoist as shown to remove steering support assembly.

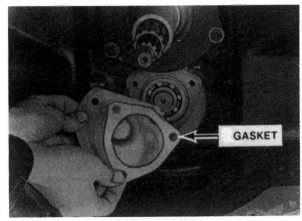

GASKET

Fig. 90—Unbolt and remove pto driven shaft end cover and metal gasket.

To disassemble the removed Power Shift unit, refer to Fig. 95 and proceed as follows: Clamp clutch shaft (28) in a soft jawed vise. Remove snap ring (7), then using a suitable puller, remove clutch assembly (8 through 19) and (33 through 37). Remove Woodruff key (29) from clutch shaft (28) and slide low drive gear (30) with sprague clutch (31) and oil seal (32). Remove oil seal and sprague clutch with brass cup from low drive gear. Remove snap ring (26) and slide housing (21) from shaft. Unlock and remove the two seal rings (23), remove snap ring (24), press bearing (25) from shaft and remove thrust washer (27).

Place clutch assembly in a press and compress reduction side clutch plates, then remove retaining ring (33). Remove back plates (34 and 37) clutch discs (35) and separator plates (36). Turn clutch cup (16) over and pry out retaining ring (8). Remove back plates (9 and 12), clutch discs (10) and separator plates (11). Using an Allen wrench, remove pressure plate bolts (19) and nuts. Remove pressure plate (18), Belleville washers (17), piston (13) and three washers from clutch cup. Remove "O" rings (14 and 15) from piston.

Clean and inspect all parts and renew any showing excessive wear or other damage. When reassembling, use all new "O" rings, seal rings and oil seals and reverse disassembly procedure keeping the following points in mind: Before installing new friction discs (10 or 35), soak discs in clean Hy-Tran Plus fluid for three minutes. Using a micrometer, measure thickness of low clutch components; retaining ring (33), back plates (34 and 37), three friction discs (35), two separator plates (36) and pressure plate (18). Measure the stack in four places. The thickest measurement will be dimension "A". Correct thickness for dimension "A" is 1.042-1.055 inches (26.466-26.797 mm). Refer to chart in Fig. 96 and select shim as required to obtain dimension "A". Shims are available in thicknesses of 0.014 inch (0.355 mm), 0.024 inch (0.609 mm) and 0.036 inch (0.914 mm).

Use petroleum jelly to stick the three washers in piston side of housing. Lubricate new "O" rings (14 and 15—Fig. 95) on piston with petroleum jelly and install piston in clutch cup. Turn clutch cup (16) over and install the three Belleville washers (17) so concave side faces outward. Aligning the holes, install

Fig. 91—With clutch release sleeve carrier removed, withdraw pto drive shaft and gear.

Fig. 93—View of snap ring and splined gear installed on front of direct drive gear.

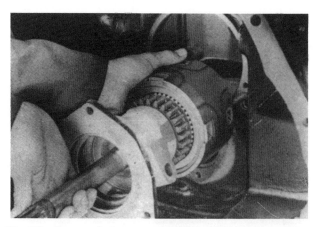

Fig. 92—Removal of two speed Power Shift unit from front of speed gearbox.

Fig. 94—Remove direct drive gear assembly as shown.

clutch pressure plate (18). Apply Loctite 270 on threads of bolts (19) and install bolts and new nuts. Tighten bolts to a torque of 5-6 ft.-lbs. (7-8 N·m). Use a center punch and stake the nuts. Install the predetermined shim, piston back plate (37), friction disc (35), separator plate (36), friction disc (35), separator plate (36), friction disc (35) and clutch back plate (34). Align notches in clutch friction discs, compress clutch plates and install retaining ring (33).

Turn clutch cup over and measure the distance from retaining ring groove to face of piston. Record as dimension "A". Using a micrometer, measure thickness of direct drive (high) clutch components; retaining ring (8), back plates (9 and 12), two separator plates (11) and three friction discs (10). Measure the stack in four places. The thickest measurement will be dimension "B". Subtract dimension "B" plus 0.035 inch (0.899 mm) from dimension "A". The result will be dimension "C". Refer to chart in Fig. 97 and select correct shim or shims as required to obtain a correct dimension "C" of 0-0.015 inch (0-0.381 mm). Install the predetermined shim, piston back plate (12—Fig. 95), friction disc (10), separator plate

(11), friction disc (10), separator plate (11), friction disc 10), clutch back plate (9) and retaining ring (8). Check to see that direct drive clutch plates are free to rotate and that minimum clearance between retaining ring (8) and back plate (9) is 0.030-0.040 inch (0.762-1.016 mm). Place clutch in a press and compress Belleville washers until the direct drive clutch plates are just locked. Then, check to see that low clutch plates are free to rotate and that minimum

Dimension A	Shim required
1.055 to 1.042 inch (26.797 to 26.466 mm)	No shim required
1.041 to 1.027 inch (26.441 to 26.085 mm)	1 x 0.014 inch (1 x 0.356 mm)
1.026 to 1.016 inch (26.060 to 25.806 mm)	1 x 0.024 inch (1 x 0.610 mm)

Fig. 96—Chart used in selecting shim to adjust correct thickness of low clutch components. Refer to text.

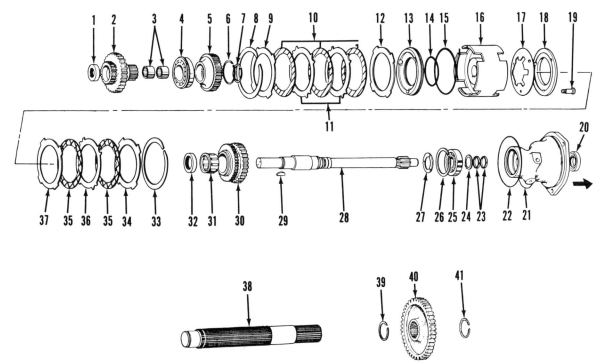

Fig. 95—Exploded view of the two speed Power Shift unit available on all models.

1. Needle bearing
2. Direct drive gear
3. Needle bearings
4. Ball bearing
5. Splined gear
6. Snap ring
7. Snap ring
8. Retaining ring
9. Clutch back plate
10. Clutch discs (3)
11. Separator plates (2)
12. Piston back plate
13. Direct drive piston
14. "O" ring
15. "O" ring
16. Clutch cup
17. Belleville washer (3)
18. Pressure plate
19. Bolt (3)
20. Oil seal
21. Housing
22. "O" ring
23. Seal rings
24. Snap ring
25. Ball bearing
26. Snap ring
27. Thrust washer
28. Clutch shaft
29. Woodruff key
30. Low drive gear
31. Sprague clutch
32. Oil seal
33. Retaining ring
34. Clutch back plate
35. Clutch discs (3)
36. Separator plates (2)
37. Piston back plate
38. Countershaft
39. Snap ring
40. Low driven gear
41. Snap ring

clearance between retaining ring (33) and clutch back plate (34) is 0.030-0.040 inch (0.762-1.016 mm). If not, recheck shimming procedures.

Install sprague clutch (31) in low drive gear (30) and install the brass cup ring. Install oil seal (32) until seal

Dimension C	Shim required
0 to 0.015 inch (0 to 0.381 mm)	No shim required
0.015 to 0.025 inch (0.381 to 0.635 mm)	0.014 inch (0.355 mm)
0.026 to 0.035 inch (0.660 to 0.899 mm)	0.024 inch (0.609 mm)
0.036 to 0.045 inch (0.914 to 1.143 mm)	0.036 inch (0.914 mm)
0.046 to 0.055 inch (1.168 to 1.397 mm)	0.014 inch + 0.036 inch (0.355 mm + 0.914 mm)
0.056 to 0.065 inch (1.422 to 1.651 mm)	0.024 inch + 0.036 inch (0.609 mm + 0.914mm)
0.066 to 0.080 inch (1.676 to 2.032 mm)	0.036 inch x 2 (0.914 mm x 2)

Fig. 98—Low drive gear and sprague clutch should rotate freely in clockwise direction (viewed from front) and lock to shaft when rotated in counterclockwise direction.

is 0.050 inch (1.27 mm) below gear face. When low drive gear and sprague clutch assembly is installed on shaft, make certain that gear will rotate clockwise (viewed from front) and will lock to shaft in opposite direction. See Fig. 98.

Reinstall speed gearbox to engine. Apply Loctite 515 gasket eliminator to mounting face of range section, then reconnect speed gearbox to range section. Tighten bolts securing speed gearbox to engine and speed gearbox to range section to a torque of 80 ft.-lbs. (108 N·m).

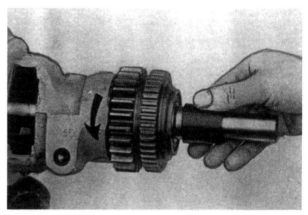

Fig. 97—Chart used in selecting shim to adjust correct thickness of direct drive (high) clutch components. Refer to text.

FORWARD-REVERSE TRANSMISSION

OPERATION

Models So Equipped

105. All models except Model 885, may be equipped with a Forward-Reverse shuttle transmission located in front of the speed transmission. The Forward-Reverse unit is controlled by a lever on left side of instrument panel. This lever is linked to the hydraulic spool in the Forward-Reverse clutch housing. Movement of this lever engages one clutch and releases the other. To prevent damage to transmission, control lever cannot be moved until engine clutch release pedal is depressed. The range transmission, on models equipped with the Forward-Reverse unit, does not have a reverse; only a high and low. This gives the tractor a total of eight forward speeds and eight reverse speeds.

TROUBLE SHOOTING

Models So Equipped

106. Some of the problems which may occur during operating of the Forward-Reverse transmission and their possible causes are as follows:

1. No drive through Forward-Reverse transmission. Could be caused by:
   a. Mechanical linkage disconnected.
   b. Low oil pressure.
   c. Regulator valve stuck open.
2. Slipping Forward-Reverse clutches. Could be caused by:
   a. Faulty regulator valve.
   b. Damaged piston "O" ring seals.
   c. Damaged or excessively worn clutch plates.

REGULATOR VALVE

Models So Equipped

107. R&R AND OVERHAUL. To remove the bottom core plate (8—Fig. 99) and regulator valve (2), first remove drain plug and drain transmission fluid. Disconnect battery cables.

> NOTE: Bottom core plate and regulator assembly can be removed without removing the speed gearbox. All other service on the Forward-Reverse transmission requires removal of complete speed gearbox.

On models equipped with front drive axle, unbolt and remove drive shaft shield and front drive shaft. Then, on all models, unbolt and remove bottom core plate and regulator assembly. Remove inlet tube (5). Pressure tubes (4) will remain in speed gearbox.

Remove plug (1—Fig. 100) with "O" ring (2), shim (3), pin (4), spring (5) and valve (6). Remove plug (9) and "O" ring (8). Unbolt and remove regulator valve body (7), then remove gasket (11—Fig. 99), separator plate (10) and gasket (9) from core plate (8).

Clean and inspect all parts for excessive wear or other damage and renew as necessary. Valve body (7—Fig. 100) is not serviced separately. Spring (5) should have a free length of 1.86 inch (47.4 mm). Use new gaskets and "O" rings and reassemble and reinstall by reversing disassembly and removal procedures.

108. PRESSURE TEST AND ADJUST. To check the oil pressure to the clutches, remove plug from test port in bottom of the regulator valve core plate. Connect a hose and a 0-500 psi (0-3500 kPa) test gage to the port. Warm transmission oil in tractor to a temperature of at least 100° F (38° C). Then, with engine operating at low idle speed, move control lever

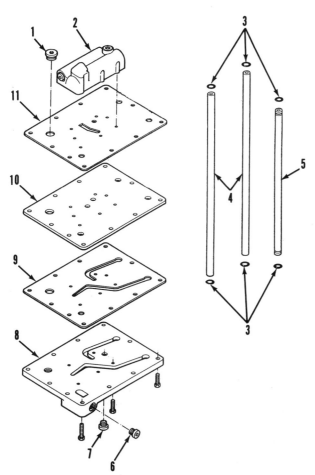

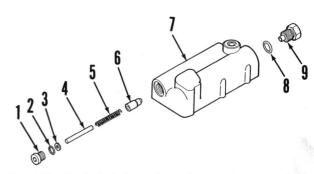

Fig. 100—Exploded view of regulator valve assembly.

1. Plug	
2. "O" ring	6. Valve
3. Shim	7. Body
4. Pin	8. "O" ring
5. Spring	9. Plug

Fig. 99—View of regulator valve and bottom core plate and relative components used on models equipped with Forward-Reverse transmission.

1. Plug	7. Plug
2. Regulator valve	8. Bottom core
3. "O" rings	plate
4. Pressure tubes	9. Gasket
5. Oil inlet tube	10. Separator plate
6. Plug	11. Gasket

Fig. 101—View showing pressure test gage installed at test port in bottom core plate. Refer to text.

to select forward direction. Record pressure gage reading. Move control to select reverse direction and record gage reading. Stop engine. Pressure readings in both forward and reverse must be 300-320 psi (2070-2210 kPa). A low reading for one direction only would indicate damaged piston seal rings for that clutch. If readings are low in both forward and reverse, add shims (3—Fig. 100) in regulator as required. Adding one shim will increase pressure by 20-30 psi (138-207 kPa). If a total of more than four shims are required, renew spring (5).

FORWARD-REVERSE UNIT

Models So Equipped

109. R&R AND OVERHAUL. To remove the Forward-Reverse unit, speed gearbox must be removed from tractor. To remove the speed gearbox, first remove cab and fuel tanks or fenders, tank and platform, as equipped, as outlined in paragraph 145 or 144. Remove plugs and drain fluid from speed gearbox and rear frame. Remove front drive axle drive shaft and shield if so equipped. Unbolt and remove pto clutch cover from bottom of rear main frame. Drive roll pin from pto clutch and shaft. Unbolt lower pto shaft cover or seal housing, as equipped, and remove with gasket from rear of tractor. Have helper support the pto clutch. Withdraw lower pto shaft and remove pto clutch. Unbolt and remove bottom core plate and regulator valve as outlined in paragraph 107. Remove oil inlet tube from Forward-Reverse clutch housing. Disconnect 1st-2nd and 3rd-4th selector rods from gear selector levers. Disconnect all interfering hydraulic lines and wiring. Using a sling and hoist (Fig. 102), lift off steering support assembly. Remove right and left brake pipes. Remove steering supply and return pipes and oil cooler supply and return pipes. Plug or cap all openings immediately to prevent foreign material from entering systems. Disconnect and remove the Forward-Reverse lockout mechanism from speed selector cover.

Place special tool (Churchill #MS 2700) or equivalent, under tractor. See Fig. 89A. Adjust splitting tool to support tractor under speed gearbox and rear main frame. Place wooden wedges between front axle and front support to prevent tipping. Unbolt range section (rear frame) from speed gearbox. After making certain that all wires and hydraulic lines are disconnected, roll engine and speed gearbox assembly forward from range section.

Attach a sling and hoist to speed gearbox and block up securely under engine oil pan. Unbolt clutch housing from engine, then separate speed gearbox from engine. If available, attach speed gearbox assembly in a transmission stand.

Remove clutch release bearing and sleeve. Drive roll pins from release fork, withdraw release shaft and remove fork. Unbolt and remove pto driven shaft end cover and metal gasket as shown in Fig. 103. Remove snap ring from front end of pto driven shaft. Working through bottom core plate opening, remove snap ring from pto driven gear. Drive the driven shaft rearward out of bearing and remove bearing. Withdraw the driven shaft and have helper remove gear as shaft is removed. Unbolt and remove clutch release sleeve carrier, then withdraw pto drive shaft and gear as shown in Fig. 104. Remove snap ring from front end of countershaft and remove the reverse gear from countershaft.

Drive expansion plug from front of speed gearbox. Drive out the two roll pins from reverse idler shaft. Install a 1/4 × 8 inch NC bolt through the expansion plug hole and into reverse idler shaft. Pull out the idler shaft (Fig. 105) and remove the reverse idler gear. Unpin the Forward-Reverse actuating lever from speed selector cover, then turn lever clockwise to disengage lever from spool valve. Refer to paragraph 110 and remove speed selector cover, selector shafts and

Fig. 102—Attach sling and hoist as shown to remove steering support assembly.

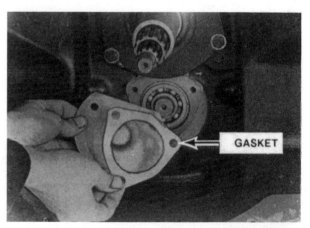

Fig. 103—Unbolt and remove pto driven shaft end cover and metal gasket.

forks. Remove the Forward-Reverse unit from front of speed gearbox. See Fig. 106. Remove pressure tubes (4—Fig. 99) and "O" rings (3).

To disassemble the removed Forward-Reverse unit, refer to Fig. 107 and proceed as follows: Clamp clutch shaft (37) in a soft jawed vise and remove shift spool (17) from housing (18). Remove retaining ring (32) and

oil deflector (33). Using a suitable puller, remove Forward-Reverse clutch unit (34). Remove Woodruff key (36), then remove drive gear (35). Remove snap ring (23) and slide housing (18) from clutch shaft. Unlock and remove the three seal rings (20) and remove retaining ring (21). Press shaft from bearing (32) and remove thrust washer (24).

To disassemble the clutch unit (34), place unit in a press (either end up) and apply pressure to clutch

Fig. 104—With clutch release sleeve carrier removed, withdraw pto drive shaft and gear.

Fig. 105—Using 1/4 x 8 inch NC bolt, withdraw idler shaft and remove reverse idler gear.

Fig. 106—Remove Forward-Reverse unit from front of speed gearbox.

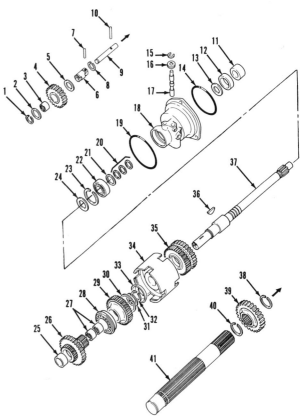

Fig. 107—Exploded view of Forward-Reverse transmission. Items (25 through 31) are located in speed gearbox.

1. Snap ring	22. Ball bearing
2. Washer	23. Snap ring
3. Needle bearing	24. Thrust washer
4. Idler gear	25. Needle bearing
5. Washer	26. Forward drive
6. Sleeve	gear & hub
7. Roll pin	27. Needle bearings
8. Snap ring	28. Ball bearing
9. Shaft	29. Clutch disc carrier
10. Roll pin	30. Snap ring
11. Oil seal	31. Retaining ring
12. Pto shaft seal	32. Retaining ring
13. Clutch shaft seal	33. Oil deflector
14. "O" ring	34. Clutch assy.
15. Retaining ring	35. Drive gear
16. Collar	36. Woodruff key
17. Forward-Reverse	37. Clutch shaft
shift spool	38. Snap ring
18. Housing	39. Countershaft
19. "O" ring	reverse gear
20. Seal rings	40. Snap ring
21. Retaining ring	41. Countershaft

plates, then remove retaining ring (1—Fig. 108). Remove clutch back plate (2), friction discs (3), separator plates (4), piston back plate (5) and shim (6). Turn clutch cup (11) over and compress clutch plates, then remove retaining ring (22). Remove clutch back plate (21), friction discs (20), separator plates (19), piston back plate (18) and shim (17). Using an Allen wrench, remove bolts (16) and nuts (8). Bump clutch cup (11) on a bench and remove piston (7) and spacers (15). Turn clutch cup over and bump out piston (14). Remove "O" rings (9, 10, 12 and 13) from pistons.

Clean and inspect all parts and renew any showing excessive wear or other damage. When reassembling, use all new "O" rings, seal rings, gaskets and oil seals. Before installing any new friction discs (3 or 20), soak new discs in clean Hy-Tran Plus fluid for three minutes. When installing pistons, tighten new bolts (16) and nuts (8) to a torque of 6 ft.-lbs. (8 N·m), then stake the nuts. Reassemble clutches in clutch cup but install one 0.024 inch (0.61 mm) shim (6). Do not install any shim (17) at this time. Install clutch assembly in a press with the side with shim (6) at the top. Apply a load of 22-30 lbs. (98-133 N) and using a feeler gage, measure the clearance between clutch back plate (2) and retaining ring (1). Clearance should be between 0.044 and 0.060 inch (1.117 to 1.524 mm). If the clearance is greater than 0.060 inch (1.524 mm), shims must be added. If clearance measures less than 0.044 inch (1.117 mm), a thinner shim must be installed. When correct clearance is obtained, remove and measure thickness of shim and divide by 2, equally as possible. For example, if shim measures 0.024 inch (0.61 mm), install one 0.014 inch (0.355 mm) thick shim on each clutch half. Shims are available in thicknesses of 0.014, 0.024 and 0.036 inch (0.355, 0.61 and 0.914 mm).

Reassemble and reinstall Forward-Reverse transmission by reversing the disassembly and removal procedures, keeping the following points in mind: Apply Loctite 515 to mounting face of range section, then reconnect speed gearbox to range section. Tighten bolts securing speed gearbox to engine and speed

gearbox to range section to a torque of 80 ft.-lbs. (108 N·m).

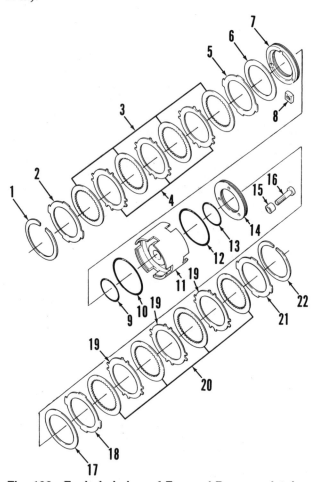

Fig. 108—Exploded view of Forward-Reverse clutches.

1. Retaining ring	12. "O" ring
2. Clutch back plate	13. "O" ring
3. Friction discs (4)	14. Piston
4. Separator plates (3)	15. Spacer (3)
5. Piston back plate	16. Bolt (3)
6. Shim	17. Shim
7. Piston	18. Piston back plate
8. Nut (3)	19. Separator plates (3)
9. "O" ring	20. Friction discs
10. "O" ring	21. Clutch back plate
11. Clutch cup	22. Retaining ring

SYNCHROMESH SPEED TRANSMISSION

The synchromesh speed transmission used on all models, is located in the rear portion of the clutch housing (speed gearbox). On models equipped with the two speed Power Shift or Forward-Reverse transmission, speed transmission is directly behind them. The speed transmission is equipped with synchronizers and therefore can be shifted with the tractor in motion by disengaging the clutch and applying a

steady and continuous pressure to the speed shift lever until the shift is complete. The speed gearbox must be removed for all service on the speed transmission.

The four speed synchromesh speed transmission in conjunction with range transmission and without the Power Shift or Forward-Reverse transmissions, provides eight forward and four reverse speeds.

R&R AND OVERHAUL

All Models

110. To remove the speed gearbox, first remove cab and fuel tanks or fenders, tank, platform and tunnel cover, as equipped, as outlined in paragraph 145 or 144. Remove drain plugs and drain fluid from speed gearbox and rear frame. If so equipped, remove front drive axle drive shaft and shield. Unbolt and remove pto clutch cover from bottom of rear main frame. Drive out roll pin (1—Fig. 109) from pto clutch and shaft. Unbolt lower pto shaft cover or seal housing, as equipped, and remove with gasket from rear of tractor. Have helper support the pto clutch. Withdraw lower pto shaft and remove pto clutch. Remove snap ring from rear of pto driven gear, then remove driven gear. Disconnect all interfering hydraulic lines and electrical wiring. Then, using a sling and hoist (Fig. 110), lift off steering support assembly. Disconnect 1st-2nd and 3rd-4th selector rods from gear selector levers. Remove right and left brake pipes. Remove steering supply and return pipes and oil cooler supply and return pipes.

NOTE: Cap or plug all openings immediately to prevent foreign material from entering systems.

Place special tool (Churchill #MS 2700) or equivalent, under tractor. See Fig. 89A. Adjust splitting tool to support tractor under speed gearbox and rear main frame. Place wooden wedges between front axle and front support to prevent tipping. Unbolt range section (rear frame) from speed gearbox. After making certain that all wires and hydraulic lines are disconnected, roll engine and speed gearbox forward from range section.

Attach a sling and hoist to speed gearbox and block up securely under engine oil pan. Unbolt clutch housing from engine, then separate speed gearbox from engine. If available, attach speed gearbox to a transmission stand.

To disassemble the removed synchromesh speed transmission, first remove the speed gearbox cover (2—Fig. 111) as follows: Disconnect brake return hose from adaptor (4) and lube hose (11) from lube tube (16). Remove snap ring (18) from lube tube (16). Make

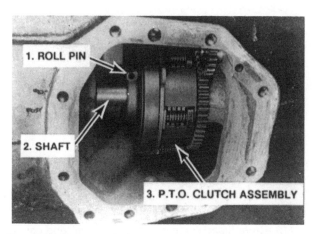

Fig. 109—Bottom view of pto clutch (cover removed).

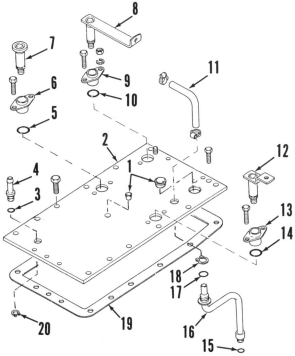

Fig. 111—View of typical speed gearbox cover and relative components.

1. Expansion plugs	
2. Cover	
3. "O" ring	11. Lube hose
4. Brake return	12. Shift lever
adaptor	(3rd-4th)
5. "O" ring	13. Housing
6. Housing	14. "O" ring
7. Interlock pivot	15. "O" ring
8. Shift lever	16. Lube tube
(1st-2nd)	17. "O" ring
9. Housing	18. Snap ring
10. "O" ring	19. Gasket
	20. Snap ring

Fig. 110—Attach sling and hoist as shown to remove steering support assembly.

certain shift levers (8 and 12) are in neutral position, then unbolt and remove cover (2). Remove lube tube (16) from gearbox. Disassembly of the cover and shift levers is obvious after examination of the unit and reference to Figs. 111 and 112.

To remove shift forks (28 and 29—Fig. 112), remove snap rings (27) from grooves in shift rail (31). Drive roll pin (30) from the rail. Push the shift rail forward and rotate rail about 1/2 turn to unseat detent balls from slots in rail. Move snap rings rearward on rail and continue moving rail forward. Cover shift fork (28) with a shop towel to prevent losing detent ball (24) and spring (25). Lift out shift fork (28). Remove shift fork (29) in same manner.

Remove clutch release bearing and sleeve. Drive roll pins from release fork, withdraw release shaft and remove fork. Unbolt and remove pto driven shaft end cover and metal gasket as shown in Fig. 113. Remove snap ring from front end of pto driven shaft. Working through bottom cover opening, remove snap ring from pto driven gear. Using a brass drift, drive the driven shaft rearward out of bearing and remove bearing. Withdraw the driven shaft and have helper remove gear as shaft is removed. Unbolt and remove clutch release sleeve carrier, then withdraw pto drive shaft and gear as shown in Fig. 114.

Before removing the mainshaft (3—Fig. 115), use a dial indicator and check end float between rear thrust washer (14) and second speed gear (13). End float must be 0.004-0.037 inch (0.102-0.94 mm). Then, measure end float between front thrust washer (14) and fourth speed gear (16). End float must be 0.0-0.076 inch (0.0-1.93 mm). If end float is excessive, new thrust washers must be installed during reassembly. Thrust washers are available in one size only.

Unbolt bearing cage (7) and slide mainshaft (3), rearward far enough for access to snap ring (15). Sep-

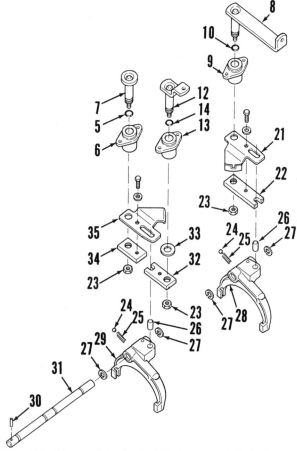

Fig. 112—View of typical shift levers, shift forks and interlocks used on synchromesh speed transmission.

5. "O" ring
6. Housing
7. Interlock pivot
8. Shift lever (1st-2nd)
9. Housing
10. "O" ring
12. Shift lever (3rd-4th)
13. Housing
14. "O" ring
21. Interlock
22. Shift arm
23. Nuts
24. Detent balls
25. Detent springs
26. Dowel pins
27. Snap rings
28. Shift fork (1st-2nd)
29. Shift fork (3rd-4th)
30. Roll pin
31. Shift rail
32. Shift arm
33. Spacer
34. Interlock support arm
35. Interlock

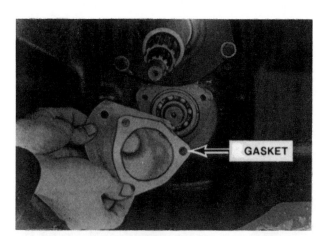

Fig. 113—Unbolt and remove pto driven shaft end cover and metal gasket.

Fig. 114—With clutch release sleeve carrier removed, withdraw pto drive shaft and gear.

arate thrust washers (14) and remove snap ring (15). Withdraw the mainshaft assembly (1 through 9) and remove thrust washers (14), second speed gear (13), synchronizer (12), hub (11) and first speed gear (10) as shaft is removed. Remove fourth speed gear (16), hub (17) and synchronizer (18). Remove spacer (9) from mainshaft. Remove snap ring (4), then remove mainshaft and bearing (5) from bearing cage (7). Remove seal rings (8) and snap ring (6), then press bearing (5) from mainshaft. If desired, remove snap ring (1) and needle bearing (2).

Remove housing (27) with "O" ring (28) and bushing (29) from front of speed gearbox. Remove transmission input (clutch) shaft (31) and bearing (30) out through rear of speed gearbox. Press bearing from shaft.

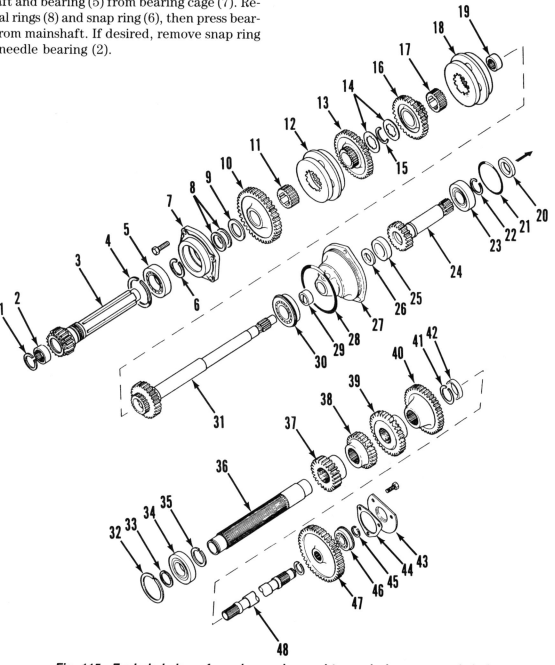

Fig. 115—Exploded view of synchromesh speed transmission gears and shafts.

1. Snap ring	11. Hub	21. "O" ring	31. Clutch shaft	40. Constant mesh gear
2. Needle bearing	12. Synchronizer	22. Snap ring	32. Snap ring	41. Snap ring
3. Mainshaft	13. 2nd speed gear	23. Bearing	33. Snap ring	42. Needle bearing
4. Snap ring	14. Thrust washers	24. Pto drive shaft	34. Bearing	43. Cover
5. Bearing	15. Snap ring	25. Oil seal	35. Snap ring	44. Metal gasket
6. Snap ring	16. 4th speed gear	26. Oil seal	36. Countershaft	45. Snap ring
7. Bearing cage	17. Hub	27. Housing	37. 1st gear	46. Bearing
8. Seal rings	18. Synchronizer	28. "O" ring	38. 2nd gear	47. Pto driven gear
9. Spacer	19. Needle bearing	29. Bushing	39. 4th gear	48. Pto driven shaft
10. 1st speed gear	20. Oil seal	30. Bearing		

To remove the countershaft (36), remove snap ring (41) from groove in shaft. Remove snap ring (32), then using a brass drift, drive countershaft rearward. Remove snap ring (41) and gears (37, 38, 39 and 40) as shaft is withdrawn. Remove snap ring (33) and press bearing (34) from countershaft. Remove needle bearing (42) from gearbox.

Clean and inspect all parts and renew any showing excessive wear or other damage. Check synchronizer friction faces and cups for excessive wear or other damage as shown in Figs. 116 and 117. Refer to Fig. 118 and press down on synchronizer ring, then release, to check centering action of springs. Synchronizers are serviced only as complete assemblies.

When reassembling, use all new "O" rings, seal rings, oil seals and gaskets and reverse disassembly procedures. When installing countershaft needle bearing, refer to Fig. 119 and install at distance shown. Apply Loctite 515 to mounting surface of range section, then reconnect speed gearbox to range section. Tighten retaining bolts to a torque of 80 ft.-lbs. (108 N·m). Refill transmission to level on dipstick with new Hy-Tran Plus fluid. Refer to Condensed Specifications for capacities.

Fig. 118—Press down on synchronizer ring, then release, to check centering action of springs.

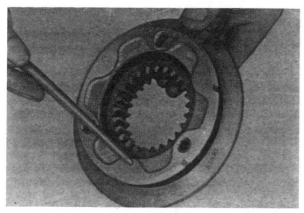

Fig. 116—Inspect synchronizer friction faces for wear or other damage.

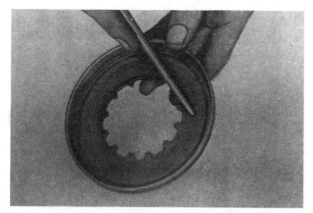

Fig. 117—Check synchronizer cups for excessive wear or other damage.

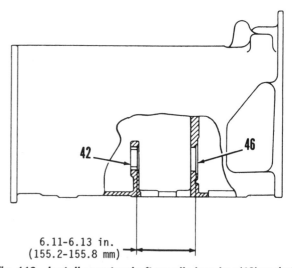

6.11-6.13 in. (155.2-155.8 mm)

Fig. 119—Install countershaft needle bearing (42) and pto ball bearing (46) as shown.

RANGE TRANSMISSION

The range transmission is located in the front portion of the tractor rear frame. The range transmission is equipped with Hi (direct drive), Lo (underdrive), neutral and reverse. On all models equipped with Forward-Reverse transmission, range transmission uses only Hi, Lo and neutral as reverse is provided by the Forward-Reverse unit. The light duty band and drum type transmission park brake is located in the range transmission. The heavy duty park brake is located on the differential ring gear.

To remove the range transmission, tractor rear main frame (range section) must be separated from speed gearbox as in paragraph 110 and the differential and the hydraulic lift housing removed.

R&R AND OVERHAUL

All Models

111. With the engine and speed gearbox assembly separated from the range section, remove hydraulic lift housing as outlined in paragraph 137 or 138. Remove the differential assembly as in paragraph 115.

Refer to Fig. 120 and remove snap rings (2). Drive out roll pins (1) and rotate shift rail (3) to disengage detent balls (4) from slots in rail. Move shift rail forward and cover shift fork (6) with a shop towel to prevent losing detent ball and spring. Rotate shift fork to remove bushing (7), then remove Lo-Reverse shift fork. Remove Hi-Neutral shift fork (16) in same manner. Removal of shift levers and shift arms is obvious after examination of the units and reference to Fig. 120.

NOTE: On models equipped with Forward-Reverse transmission, only one shift lever and shift fork is used.

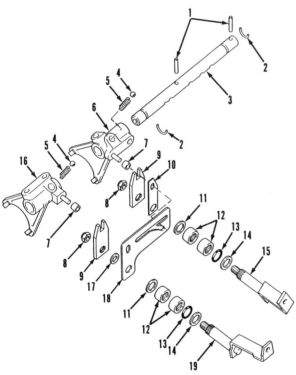

Fig. 120—Exploded view of shift levers, shift rail and shift forks used on range transmission.

1. Roll pins	
2. Snap rings	12. Needle bearings
3. Shift rail	13. "O" rings
4. Detent balls	14. Washers
5. Detent springs	15. Lo-Reverse shift
6. Lo-Reverse shift	lever
fork	16. Hi-Neutral shift
7. Bushings	fork
8. Nuts	17. Washer
9. Shift arms	18. Interlock plate
10. Spacer plate	19. Hi-Neutral shift
11. Washers	lever

112. MODELS 385-485 (TWO-WHEEL DRIVE). To remove the bevel pinion mainshaft (1—Fig. 121), loosen park brake adjustment, then unpin and remove park brake band (6). Remove snap ring (9) from groove in bevel pinion shaft. Unbolt bearing housing (3), move bevel pinion shaft rearward and remove Hi sliding collar (16). Remove Lo-Reverse gear (15), snap ring (9) and brake drum (8). Remove bevel pinion mainshaft assembly with shims (5) from transmission case. Measure and note thickness of shims if the same bevel pinion mainshaft is to be reinstalled. Using a dial indicator, measure end play of bearing housing (3) on bearings (2 and 4). If end play is more than 0.009 inch (0.229 mm), renew the bearings as follows: Remove snap ring (7) and press bearings and housing from bevel pinion mainshaft. Install new bearings (2 and 4) and housing (3) and secure with snap ring (7).

To remove the countershaft (25), remove snap ring (18) and remove snap ring (24) from its groove in countershaft. Drive a wedge into split in spacer (22). Move countershaft forward and remove snap ring (24), Lo gear (23), split spacer (22) and constant mesh gear (21) as countershaft is withdrawn. Remove snap ring (17), then press ball bearing (19) from countershaft. Pump drive gear (29) is located on pto driven shaft and rides in needle bearing (27). If necessary, countershaft needle bearing (26) and pump gear needle bearing (27) can now be removed. Note position of needle bearings before removing.

To remove the reverse idler gear (35) and shaft (32), drive out roll pin (33). Withdraw shaft (32) and remove reverse idler gear (35) with needle bearings (34 and 37) and spacer (36), then remove thrust washers (31 and 38).

Clean and inspect all parts and renew any showing excessive wear or other damage. If bevel pinion mainshaft must be renewed, bevel ring gear must also be renewed as they are available only as a matched set.

To set bevel pinion mainshaft position, install bearings (2 and 4), housing (3) and snap ring (7). The mounting distance (A—Fig. 122) from mainshaft pinion bearing housing mounting face to the center line of the differential is approximately 6.600 inches (167.64 mm). This distance may vary for each tractor. To aid in accurate installation of the pinion mainshaft, a single figure is stamped on the top face of the upper pto shaft boss as shown in Fig. 123. The single figure indicates (in inches) the variation from distance (A—Fig. 122) in steps of 0.001 inch (0.025 mm). When distance (A) is smaller, a minus sign will be shown in front of single figure. Add or subtract the single figure to the approximate distance 6.600 inches (167.64 mm). The pinion cone mounting distance (B) is marked (in inches) on the rear face of the pinion as shown in Fig. 124 and shows the distance this face must be from center line of differential. The distance will be approximately 4.410 inches (112.01 mm). Stand pinion shaft on a flat surface. See Fig. 125. Measure distance from pinion face to the bearing housing mounting face. This is distance (C—Fig. 122). Add distance (C) to the marked figure on pinion face (distance B) and subtract from the resulting (distance A). The final figure is the correct thickness of shims (5—Fig. 121) to be installed for correct pinion mainshaft position.

The balance of reassembly is the reverse of disassembly procedures.

113. MODELS 585-685-885 (TWO-WHEEL DRIVE) AND ALL MODELS (FOUR-WHEEL DRIVE). To re-

move the bevel pinion mainshaft (2—Fig. 126), remove snap ring (14) from groove in pinion shaft. Unbolt retaining plates (6) and move bevel pinion shaft rearward. Remove Hi sliding collar (16), Lo-Reverse gear (15), snap ring (14) and four-wheel drive gear (13), if so equipped. Remove pinion mainshaft assembly and shims (10) from transmission case. Measure and note thickness of shims (10) if the same bevel pinion mainshaft is to be reinstalled. Using a dial indicator, measure end play of bearing housing (8) on bearings (7 and 9). If end play is more than 0.003 inch (0.076 mm), renew the bearings as follows: Remove snap ring (12) and spacer (11), then press bearings and housing from bevel pinion mainshaft. Press new bear-

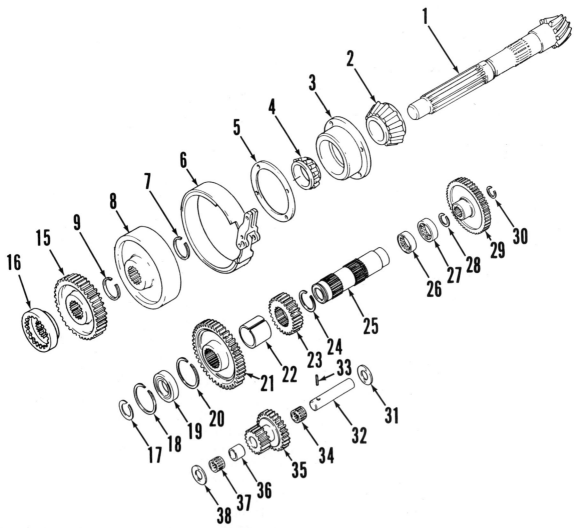

Fig. 121—Exploded view of range transmission used on Models 385 and 485 without front drive axle. Note the light duty park brake (6 and 8).

1. Bevel pinion mainshaft	8. Brake drum	20. Snap ring	26. Countershaft needle bearing	32. Shaft
2. Rear bearing	9. Snap ring	21. Constant mesh gear	27. Pump drive gear needle bearing	33. Roll pin
3. Bearing housing	15. Lo-Reverse gear			34. Needle bearing
4. Front bearing	16. Hi sliding collar	22. Split spacer	28. Snap ring	35. Reverse idler gear
5. Shim	17. Snap ring	23. Lo gear	29. Pump drive gear	
6. Brake band	18. Snap ring	24. Snap ring	30. Snap ring	36. Spacer
7. Snap ring	19. Ball bearing	25. Countershaft	31. Thrust washer	37. Needle bearing
				38. Thrust washer

ings (7 and 9) and housing (8) on bevel pinion while rotating the housing. Apply pressure until a slight preload can be felt. Install spacer (11). Snap ring (12) is available in thicknesses of 0.066, 0.069, 0.072, 0.075, 0.078, 0.081, 0.084, 0.087, 0.090 and 0.093 inch (1.67, 1.75, 1.83, 1.90, 1.98, 2.06, 2.13, 2.21, 2.28 and 2.36 mm). Select the thinnest snap ring, 0.066 inch (1.67 mm) and install in its groove. Make certain that bearing (9) and spacer (11) are tight against snap ring (12), then using a dial indicator, measure end play of housing (8) on bearings (7 and 9). Add the end play to the thickness of installed snap ring. The result is the correct thickness of snap ring to be installed. Remove the thin snap ring and install the new one. This assembly will now have a maximum end play of 0.003 inch (0.076 mm) or a maximum preload of 0.002 inch (0.050 mm).

To remove the countershaft (25), remove snap ring (18) and remove snap ring (24) from its groove in countershaft. Drive a wedge into split in spacer (22). Move countershaft forward and remove snap ring

(24), Lo gear (23), split spacer (22) and constant mesh gear (21) as countershaft is withdrawn. Remove snap ring (17), then press ball bearing from countershaft. Pump drive gear (29) is located on pto driven shaft and rides in needle bearing (27). If necessary, countershaft needle bearing and pump gear needle bearing (27) can now be removed. Note position of needle bearings before removing.

To remove the reverse idler gear (35) and shaft (32), drive out roll pin (33). Withdraw shaft (32) and remove reverse idler gear (35) with needle bearings (34 and 37) and spacer (36), then remove thrust washers (31 and 38).

Clean and inspect all parts and renew any showing excessive wear or other damage. If bevel pinion mainshaft must be renewed, bevel ring gear (1) must also be renewed as they are available only as a matched set.

To set bevel pinion mainshaft position, install bearings (7 and 9), housing (8), spacer (11) and correct snap ring (12) as outlined previously, then proceed as follows: The mounting distance (A—Fig. 122) from pinion bearing housing mounting face to the center line of the differential is approximately 6.600 inches (167.64 mm). This distance may vary for each trac-

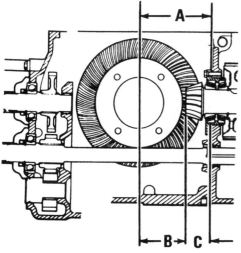

Fig. 122—View showing distances (A, B and C) used in setting pinion position.

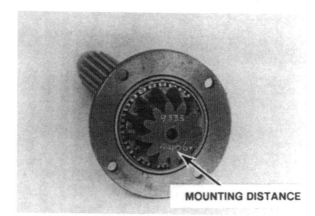

Fig. 124—Mounting distance (B—Fig. 122) is marked (in inches) on rear face of pinion. Refer to text.

Fig. 123—A single variation figure is stamped (in inches) on top face of pto upper shaft boss. Refer to text.

Fig. 125—Measure distance (C—Fig. 122) from pinion face to bearing housing mounting face. Refer to text.

tor. To aid in accurate installation of the pinion main-shaft, a single figure is stamped on the top face of the upper pto shaft boss as shown in Fig. 123. The single figure indicates (in inches) the variation from distance (A—Fig. 122) in steps of 0.001 inch (0.025 mm). When distance (A) is smaller, a minus sign will be shown in front of single figure. Add or subtract the single figure to the approximate distance 6.600 inches (167.64 mm). The pinion cone mounting distance (B) is marked (in inches) on the rear face of the pinion as shown in Fig. 127 and shows the distance this face must be from center line of differential. The distance will be approximately 4.410 inches (112.01 mm). Stand pinion shaft on a flat surface. See Fig. 128. Measure distance from pinion face to the bearing housing mounting face. This is distance (C—Fig. 122). Add distance (C) to the marked figure on pinion face (distance B) and subtract from the resulting (distance A). The final figure is the correct thickness of shims (10—Fig. 126) to be installed for correct pinion mainshaft position.

The balance of reassembly is the reverse of disassembly procedures.

TRANSFER LUBE PUMP

All Models

114. REMOVE AND REINSTALL. The transfer lube pump is located below the reverse idler gear in range gearbox. The pump furnishes oil for lubricating the speed transmission mainshaft and synchronizers and the ring gear and pinion.

To remove the transfer pump, first remove all other range transmission components as outlined in paragraphs 111, 112 and 113. Then, disconnect and

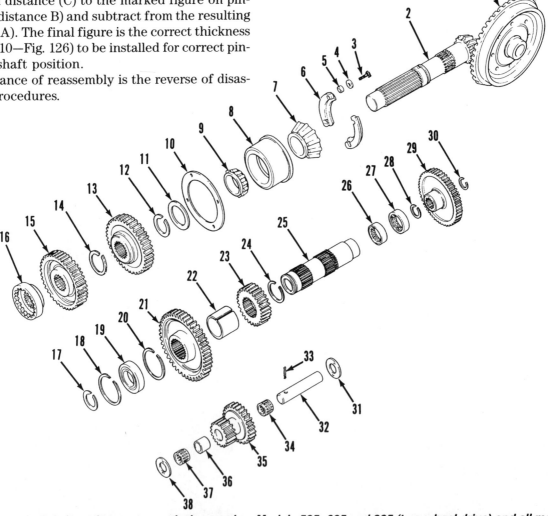

Fig. 126—Exploded view of range transmission used on Models 585, 685 and 885 (two-wheel drive) and all models with four-wheel drive. Note the heavy duty park brake drum (flange) on bevel ring gear.

1. Bevel ring gear	9. Bearing, front	17. Snap ring	31. Thrust washer
2. Bevel pinion mainshaft	10. Shim	18. Snap ring	32. Shaft
3. Cap screw	11. Spacer	19. Ball bearing	33. Roll pin
4. Washer	12. Snap ring	20. Snap ring	34. Needle bearing
5. Lock collar	13. Front-wheel drive gear	21. Constant mesh gear	35. Reverse idler gear
6. Retaining plates	14. Snap ring	22. Split spacer	36. Spacer
7. Bearing, rear	15. Lo-Reverse gear	23. Lo gear	37. Needle bearing
8. Bearing housing	16. Hi sliding collar	24. Snap ring	38. Thrust washer
		25. Countershaft	
		26. Countershaft needle bearing	
		27. Pump drive gear needle bearing	
		28. Snap ring	
		29. Pump drive gear	
		30. Snap ring	

remove lubrication tube (1—Fig. 130). Remove snap ring (8) and drive gear (7). Remove Allen set screw (5) through hole in bottom of rear main frame shown in Fig. 129. Remove pump from transmission housing. Unbolt and remove strainer (2—Fig. 130), then remove end cover (3). Clean and inspect all parts. Internal parts of pump (4) are not available separately.

Reassemble and reinstall transfer pump by reversing the disassembly and removal procedures.

Fig. 127—Mounting distance (B—Fig. 122) is marked (in inches) on rear face of pinion. Refer to text.

Fig. 128—Measure distance (C—Fig. 122) from pinion face to bearing housing mounting face. Refer to text.

Fig. 129—Remove transfer pump retaining screw through hole in bottom of rear frame.

MAIN DRIVE BEVEL GEARS AND DIFFERENTIAL

The differential is carried on tapered roller bearings. The bearing on the bevel gear side of the differential is larger than that on the opposite side and therefore, it is necessary to keep the bearing cages in the proper relationship.

The differential assembly for Models 385 and 485 has two pinions (9—Fig. 131) with a single cross shaft (12). Models 585, 685 and 885 have four pinions (9) and four stud shafts (10). The differential lock (14) is on the right side of all models.

The bevel drive pinion is the range transmission mainshaft. Refer to paragraphs 111, 112 or 113 for R&R procedure and adjustment of bevel pinion shaft. On two-wheel drive Models 385 and 485, the light duty park brake drum is on the bevel drive pinion shaft. On two-wheel drive Models 585, 685 and 885 and all four-wheel drive models, the heavy duty park brake drum is the flange on left side of bevel ring gear.

R&R AND OVERHAUL DIFFERENTIAL

All Models

115. To remove the differential assembly, first remove cab or fenders, fuel tank and seat platform as outlined in paragraph 145 or 144. Remove hydraulic lift housing as in paragraph 137 or 138 and final

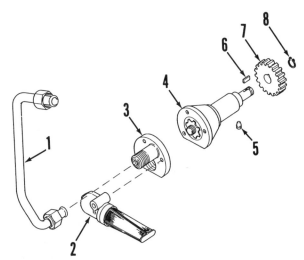

Fig. 130—Partially disassembled transfer pump and relative components.

1. Lube tube	5. Allen set screw
2. Strainer	6. Key
3. End cover	7. Gear
4. Pump	8. Snap ring

drives as in paragraph 118. Disconnect and remove bevel pinion and ring gear lubrication tube. Disconnect and remove differential lock shaft and fork. On models so equipped, disconnect and remove heavy duty park brake cam pivot and brake band hanger support. On all models, remove inner brake ring from each side. Refer to Fig. 131 and using special tool IH 7847A and a slide hammer, pull out brake pistons. Install a rope through differential carrier for a sling and attach it to a hoist. Support differential with the hoist and unbolt and remove bearing housings (18 and 22—Fig. 132) and shims (21). Identify and keep shims with their relevant bearing housing. Remove differential and heavy duty park brake band, if so equipped. See Fig. 133. Slip off the brake band.

Remove bolts (3 and 4—Fig. 132) and separate bevel ring gear (6) from differential carrier (11). Remove thrust washer (7).

> **NOTE: The four dowel bolts (3) secure the four stub shafts (10). On Models 385 and 485, single cross shaft (12) and two pinions (9) are used.**

Remove stub shafts (10) or cross shaft (12), then remove side gears (8) and differential pinions (9). Using a suitable puller, remove bearing cones (2 and 15). Remove differential lock ring (14) and springs (13). If necessary, remove bearing cups (1 and 16) from bearing housings (18 and 22).

Clean and inspect all parts and renew any showing excessive wear or other damage. If bevel ring gear is to be renewed, bevel pinion shaft (5) must also be renewed as they are available only as a matched set. Refer to paragraphs 111, 112 or 113 for installation and position adjustment of the bevel pinion shaft. Reassemble by reversing the disassembly procedure keeping the following points in mind: Thrust washer (7) has a lug which locates in a hole in bevel ring gear. Use petroleum jelly to stick thrust washer in ring gear for aid in installation to differential carrier. Use Loctite 241 on threads of bolts (3 and 4) and tighten bolts to a torque of 115-130 ft.-lbs. (156-176 N·m).

Reinstall differential and adjust carrier bearing preload as follows: Install right bearing housing (18) without shims (21) and "O" rings (19 and 20). Tighten cap screws (17) evenly to a torque of 115-130 ft.-lbs. (156-176 N·m). Install left bearing housing (22) without "O" rings (19 and 20) but with all shims (21) removed from right and left sides. Tighten left side

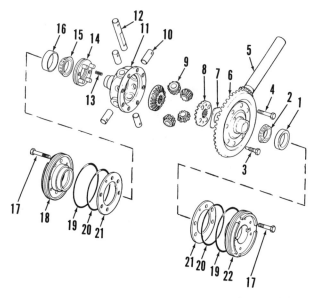

Fig. 132—Exploded view of differential assembly.

1. Bearing cup, (L.H.)
2. Bearing cone, (L.H.)
3. Dowel bolt
4. Bolt
5. Bevel pinion shaft
6. Bevel ring gear
7. Thrust washer
8. Side gears
9. Differential pinions
10. Stub shafts
11. Differential carrier
12. Cross shaft (385 & 485)
13. Spring (2)
14. Differential lock ring
15. Bearing cone, (R.H.)
16. Bearing cup, (R.H.)
17. Cap screws
18. Bearing housing, (R.H.)
19. "O" rings
20. "O" rings
21. Shims
22. Bearing housing, (L.H.)

Fig. 131—Use special tool IH7847A and slide hammer to remove brake pistons.

Fig. 133—Use a rope sling and hoist to remove differential assembly.

cap screws (17) evenly until a slight amount of end play can be felt at ring gear. Remove sling and hoist from differential. Rotate differential several turns to make sure carrier bearings are seated. Wind a cord around the differential lock ring (14) and connect a spring scale to end of cord. Make certain that range gears are in neutral position. Pull the spring scale and note the amount of pull required to rotate the differential assembly. Then, continue tightening left side cap screws evenly to a torque of 115-130 ft.-lbs. (156-176 N·m). Again, using the spring scale and cord, check amount of pull needed to rotate the differential assembly. The bearing preload is correct when a rolling pull on scale is 4.5-14.5 lbs. (2.0-6.5 kg). Add or remove shims as required to obtain correct amount of pull. One shim thickness of 0.001 inch (0.025 mm) will change amount of pull 1.0 lb. (0.64 kg). Shims are available in thicknesses of 0.003, 0.007, 0.012 and 0.030 inch (0.076, 0.170, 0.300 and 0.760 mm).

NOTE: Always support the differential assembly when removing the bearing housings.

With bearing preload set, adjust bevel gear backlash as follows: Support differential assembly and remove bearing housings (18 and 22). Divide the shims and install half on each bearing housing. Install right bearing housing (18) and tighten cap screws to a torque of 115-130 ft.-lbs. (156-176 N·m). Install left bearing housing (22) and tighten left side cap screws to a torque of 115-130 ft.-lbs. (156-176 N·m). Make sure some backlash exists between bevel ring gear (6) and bevel drive pinion (5). Using a dial indicator, check backlash. Backlash should be 0.006-0.012 inch (0.15-0.30 mm). Check backlash every 90 degrees of ring gear. Backlash must not vary more than 0.004 inch (0.1 mm) during the four readings. To increase backlash, remove shims from right side and install on left side. To decrease backlash, remove shims from left side and install on right side. Do not change total number of shims as this will affect bearing preload. Shim pack thickness must not be more than 0.071 inch (1.80 mm) on either side.

With bearing preload and backlash adjusted, remove bearing housings and install new "O" rings (19 and 20). Lubricate "O" rings with petroleum jelly. Apply Loctite 241 to bearing housing cap screws (17), then tighten cap screws to a torque of 115-130 ft.-lbs. (156-176 N·m).

Lubricate new "O" rings with petroleum jelly and install on brake pistons. Install brake pistons, then install brake inner rings. The balance of reassembly is the reverse of disassembly procedures.

DIFFERENTIAL LOCK

All tractors are equipped with a differential lock which operates on the right side of the

differential with a dowel pin type coupling locking the differential gears. It is operated with a pedal on right side of tractor.

Refer to Fig. 134 for an exploded view of differential lock linkage and to items (13 and 14— Fig. 132) for differential lock.

All Models

116. R&R AND OVERHAUL. Removal and overhaul of the differential lock requires removal and separation of the differential assembly as outlined in paragraph 115.

Overhaul of the differential lock is obvious after examination of the unit. To adjust the linkage, refer to Fig. 134 and disconnect link (9). Rotate shaft and lever (12) forward until fork (17) is fully bottomed on actuating pin (16). At this time differential lock should be fully engaged. Rotate lever and shaft (12) rearward until differential lock is fully disengaged. With pedal up, adjust length of adjusting link (9) until link will align with hole in lever. Reconnect link and actuate pedal to see that differential lock will engage and disengage properly. Adjust nuts (1), if necessary, to make sure spring (3) fully raises pedal to disengaged position. Overall length of adjusted spring (3) should be approximately 9 inches (229 mm).

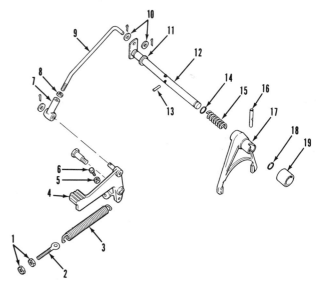

Fig. 134—Exploded view of the differential lock.

1. Adjusting nuts	10. Washers
2. Eye bolt	11. Setting washers
3. Pedal return	12. Shaft & lever
spring	13. Roll pin
4. Pedal	14. "O" ring
5. Jam nut	15. Spring
6. Stop bolt	16. Actuator pin
7. Connector	17. Fork
8. Jam nut	18. "O" ring
9. Link	19. Bushing

PARK BRAKE

All Models

117. A light duty park brake is used on Models 385 and 485 equipped with two-wheel drive. Park brake drum is located on the bevel drive pinion (range transmission mainshaft). Refer to paragraph 112 of the RANGE TRANSMISSION section for R&R procedures.

A heavy duty park brake is used on Models 585, 685 and 885 equipped with two-wheel drive and all models equipped with four-wheel drive. The park brake drum is the flange on the bevel ring gear. Refer to paragraph 115 for R&R procedures.

To adjust the light duty park brake, refer to Fig. 135 and turn adjusting nut clockwise, as viewed from front, until a pull of 10 lbs. (44.48 N) will apply park brake two or three notches on the ratchet. A pull of 10 lbs. (44.48 N) must not pull the lever past the fourth notch. When no further adjustment can be made at the adjusting nut, a new brake band must be installed.

To adjust the heavy duty park brake, refer to Fig. 136 and loosen locknut on lower clevis. Disconnect lower clevis from actuating arm. Rotate the clevis to adjust the rod length. Shorten the rod length to tighten the brake. Reconnect linkage and check park brake engagement. A pull of 70 lbs. (311 N) must not pull brake hand lever past the fourth notch on ratchet. When no further adjustment can be made on clevis, a new brake band must be installed.

On all models, place hand lever so the pawl is engaged in the first notch on ratchet. Adjust nuts on brake warning light switch so that the switch plunger is just in contact with park brake handle. Move park brake handle fully downward (disengaged po-

sition. Switch plunger should be pushed in 0.216-0.275 inch (5-7 mm) and brake warning light must be off.

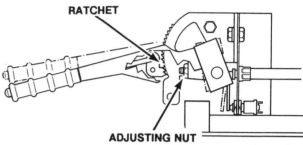

Fig. 135—View showing adjustment point on light duty park brake linkage.

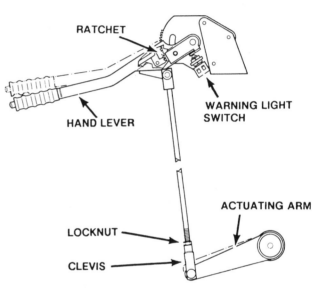

Fig. 136—View showing adjustment point on heavy duty park brake linkage.

FINAL DRIVE

The final drive assemblies consist of the rear axle, planetary unit and outer brake disc ring and can be removed from the tractor as a unit.

All Models

118. REMOVE AND REINSTALL. To remove either final drive, remove drain plug and drain housing. On models without cab, remove fenders, tank and seat platform as in paragraph 144. On models equipped with cab, raise cab and unbolt and remove the cab rear mounting brackets. Raise rear of tractor and remove rear wheel and tire assembly. Unpin and remove lower hitch link. Remove two opposite

cap screws (7—Fig. 137) and install two guide studs. Attach a hoist and sling to axle housing, then unbolt and remove final drive assembly.

Reinstall by reversing the removal procedure. Use new gasket (8) on early models or apply Loctite 515 gasket eliminator to mounting faces on later models. Install and tighten cap screws (7) to a torque of 220-250 ft.-lbs. (298-339 N·m).

119. OVERHAUL PLANETARY CARRIER. With final drive assembly removed as outlined in paragraph 118, proceed as follows: Remove outer brake disc (12—Fig. 137) and cap screw (23) securing planetary unit to axle shaft (1).

NOTE: When removing planetary unit, check for shims (21) between retainer (22) and axle shaft (1). Keep the shims together.

With planetary carrier removed from final drive assembly, drive roll pins (18) into the planet pins (19) from outside of carrier (20). Then, drive planet pins out of carrier and remove each planet gear (17), bearings (16), shims (15) and retainer (22). Remove roll pins from planet pins for reassembly. Clean and inspect gears and bearings for excessive wear or other damage and renew as necessary.

When reassembling, it will be necessary to determine shims (15) needed for preload of planet gear bearings (16) as follows: Using two flat washers made up to dimensions shown in Fig. 138, and a 1/2 × 3

inch cap screw and nuts, proceed as follows: Secure cap screw in a vise with threaded end pointing upward. Lubricate bearings (16) with Hy-Tran Plus oil and position over the cap screw with bearing and washer on each side of gear with washer flats in alignment. Tighten nut to a torque of 10 in.-lbs. (1.23 N·m) while rotating gear. Using a micrometer, measure distance across bearing inner races as shown in Fig. 139. With this measurement, refer to chart in Fig. 140 for the required shims (15—Fig. 137) to be installed for correct bearing preload. Keep each shim pack and gear together after correct shims have been selected.

When reassembling, install one planet gear with bearings and its shim pack into carrier (20) and secure in place with planet pin (19) and roll pin (18). Then, place retainer (22) in carrier and install the other two planet gears, bearings and their selected shim packs.

120. OVERHAUL REAR AXLE AND RING GEAR. If it is necessary to remove the ring gear (11—Fig. 137), remove planet carrier assembly as outlined in paragraph 119. Using a suitable puller, remove ring

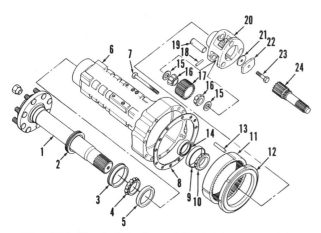

Fig. 137—Exploded view of final drive assembly.

1. Axle shaft		13. Dowel pin	
2. Oil seal		14. Oil seal	
3. Wear sleeve		15. Shim	
4. Bearing cone		16. Bearing cone	
5. Bearing cup		17. Planet gear	
6. Axle housing		18. Roll pin	
7. Cap screw		19. Planet pin	
8. Gasket (early)		20. Planetary carrier	
9. Bearing cup		21. Shim	
10. Bearing cone		22. Retainer	
11. Ring gear		23. Cap screw	
12. Brake disc		24. Planetary shaft	

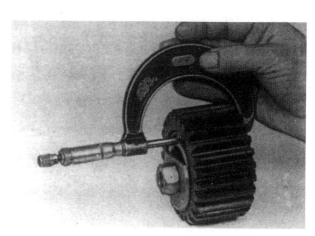

Fig. 139—Measure distance across bearing inner races as shown.

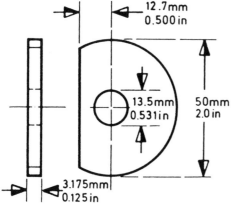

Fig. 138—Make up two flat washers to the dimensions shown.

TRACTORS	FIGURE OBTAINED	SHIMS REQUIRED
385 & 485	1.692-1.698 in. (42.98-43.13 mm)	0.056 in. (1.42 mm)
	1.699-1.705 in. (43.15-43.31 mm)	0.049 in. (1.24 mm)
	1.706-1.712 in. (43.33-43.48 mm)	0.042 in. (1.07 mm)
585,685 & 885	2.113-2.117 in. (53.66-53.76 mm)	0.056 in. (1.42 mm)
	2.118-2.122 in. (53.79-53.89 mm)	0.049 in. (1.24 mm)
	2.123-2.127 in. (53.91-54.02 mm)	0.042 in. (1.07 mm)

Fig. 140—Chart used to determine thickness of shims needed for planet gear bearings. Refer to text.

gear (11) and dowels (13). Then, using a pusher arrangement, push on inner end of axle shaft to remove axle and bearings.

Inspect bearings and seals for damage or wear and renew as necessary.

To reassemble, install axle bearing cups (5 and 9) and inner oil seal (14) in axle housing. If necessary, install wear sleeve (3) in outer end of axle housing. Install outer oil seal (2) and outer bearing cone (5) on axle shaft.

NOTE: On Row Crop models, outer oil seal can be installed after axle shaft is installed in housing.

Pack inner and outer bearing cones with multipurpose grease. Install the axle into the housing and drive the inner bearing cone (10) on axle until the planetary carrier assembly can be installed. Install and tighten cap screw (23) to draw axle shaft into inner bearing cone. Remove the planetary assembly and install ring gear (11) and secure with dowels (13).

Install a small piece of lead at end of axle shaft. Install planetary assembly and bolt (23). Lock planetary gear and ring gear by installing a rod between the teeth. Then, tighten cap screw (23) until there is 0.001-0.010 inch (0.025-0.254 mm) axle end play. Check with dial indicator. Unbolt and remove the planetary assembly. Remove the lead and measure the thickness. Select shims (21) equal to that amount. Shims are available in thicknesses of 0.007, 0.012 and 0.029 inch (0.178, 0.305 and 0.737 mm). Use petroleum jelly and stick shims to end of axle shaft. Install planetary assembly and tighten the cap screw to a torque of 220-250 ft.-lbs. (298-339 N·m). Install outer brake disc (12).

BRAKES

Brakes on all models, are self equalizing hydraulic actuated wet type single disc brakes. Brakes are located on the differential output shafts (planetary drive shafts) and are accessible after removing final drive units as outlined in paragraph 118. Return hydraulic fluid from the oil cooler maintains a full master cylinder by flowing through a brake reservoir. Brake operation can be accomplished with engine inoperative because of the reservoir which keeps master cylinders filled. Also in the brake system is an equalizer line between the master cylinders. This line permits equal flow to both brake pistons when both brake pedals are depressed. If one brake is applied, the equalizer line does not function.

Service (foot) brakes MUST NOT be used for parking or any other stationary job which re-quires the tractor to be held in position. Even a small amount of fluid leakage would result in brakes loosening and severe damage to equipment or injury to personnel could result. USE PARK BRAKE when parking tractor. Refer to paragraph 117 for park brake adjustment.

BRAKE ADJUSTMENT

All Models

121. The only adjustment that can be made is the brake pedal height.

To adjust the pedal height on models without cab, remove right rear side panel and disengage the pedal lock. Refer to Fig. 141 and loosen locknut on left pedal adjuster. Turn eccentric pivot bolt to set left pedal height to 8-1/4 inches (210 mm), measured at

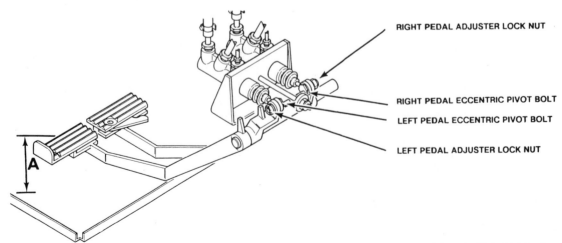

RIGHT PEDAL ADJUSTER LOCK NUT

RIGHT PEDAL ECCENTRIC PIVOT BOLT

LEFT PEDAL ECCENTRIC PIVOT BOLT

LEFT PEDAL ADJUSTER LOCK NUT

Fig. 141—View showing brake pedal height adjustment on all models not equipped with cab. Height (A) should be 8-1/4 inches (210 mm).

right angle from foot plate. Tighten locknut. Loosen right pedal adjuster locknut and turn pivot bolt to set right pedal 8-1/4 inches (210 mm) from foot plate. Check to see that pedal lock will engage easily. If not, adjust right pedal until lock engages easily. Tighten locknut. Install right side panel.

To adjust the pedal height on models equipped with cab, remove hood and grille assembly. Remove rear side panels and instrument panel front cover. Disconnect battery cables, then unbolt and raise instrument panel as necessary. Refer to Fig. 142 and loosen locknut at clevis on both pedals. Remove clevis pins. Loosen locknuts on stop bolts and turn stop bolts counterclockwise several turns. Pull both clevises free from pedals and adjust each clevis to obtain the correct pedal height of 7-7/8 to 8-1/8 inches (199-205 mm). Install clevis pins. Check to see that both pedals are of equal height and that pedal lock will engage easily. If not, readjust left pedal height to correct. Press each pedal down 1/4 inch (6 mm) and check to see that there is a slight amount of free play at the push rods. If not, turn clevis 1/2 turn counterclockwise. Tighten clevis locknuts. Turn pedal stop bolts clockwise until they contact pedals, then turn clockwise 1/4 of a turn and tighten locknuts.

BLEED BRAKES

All Models

122. To bleed air from brakes, start tractor and let it run for a few minutes to ensure the brake lines are filled. Keep engine operating while bleeding brakes. Clamp brake return hose to the transmission to prevent return oil flow. Attach two plastic hoses 1/4 inch (6 mm) ID × 30 inches (76.2 cm) long over brake bleeder fittings. Run opposite end of hoses into the

transmission filler plug hole. Loosen both air removal screws one full turn. Start engine and operate at low idle speed. Make sure brake pedals are latched together. Press pedals down quickly and release slowly. Continue until fluid with no air bubbles flows from hoses.

Tighten left air remove screw and unlatch pedals. Hold right pedal half way down. Press left pedal down quickly and release slowly until fluid with no air bubbles flows from right air removal hose. Tighten right air removal screw and loosen left screw one full turn. Hold left pedal half way down. Press right pedal down quickly and release slowly until fluid with no air bubbles flows from left air removal hose. Tighten left air removal screw. Remove clamp from brake return hose.

Latch pedals together and check operation of brakes. If brakes feel soft, repeat bleeding operation. Remove bleed hoses and install transmission filler plug.

BRAKE ASSEMBLIES

All Models

123. R&R AND OVERHAUL. Removal of either brake is accomplished by removing final drive as outlined in paragraph 118. Refer to Fig. 143 and remove planetary drive shaft (3), brake disc (2) and inner brake ring (1). Use special tool IH 7847A and a slide hammer to remove brake piston as shown in Fig. 144. Remove "O" ring from brake piston. Unbolt and remove differential bearing housing (4—Fig. 143), being careful not to lose shims. Apply petroleum jelly to grooves in bearing housing, then install new "O" rings. Reinstall brake housing and shims and using Loctite 241 on cap screw threads, tighten cap screws

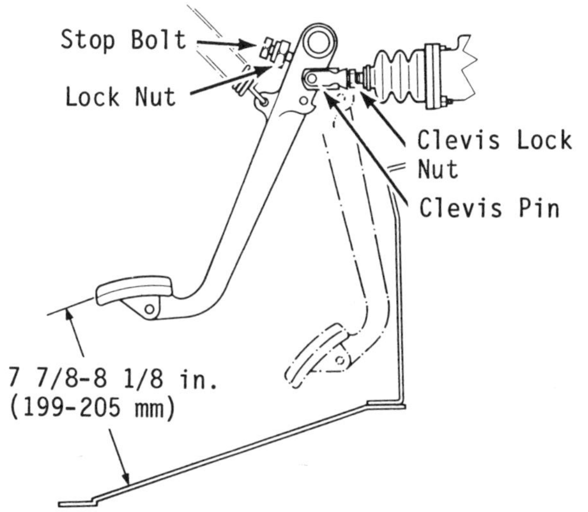

Stop Bolt

Lock Nut

Clevis Lock Nut

Clevis Pin

7 7/8-8 1/8 in. (199-205 mm)

Fig. 142—View showing brake pedal height adjustment on all models equipped with cab.

Fig. 143—Planetary drive shaft and brake disc being removed from differential housing.

1. Inner brake ring
2. Brake disc
3. Planetary drive shaft
4. Bearing housing

to a torque of 115-130 ft.-lbs. (156-176 N·m). Check condition of piston, brake disc and brake rings and renew as necessary.

When reassembling, install new "O" ring on brake piston and lubricate "O" ring with petroleum jelly. Install piston in rear frame, being careful not to damage the "O" ring.

Reassemble by reversing the disassembly procedure. When assembly is complete, check pedal free height and adjust as outlined in paragraph 121, then start engine and bleed brakes as in paragraph 122.

BRAKE MASTER CYLINDERS

All Models

124. R&R AND OVERHAUL. To remove brake master cylinders, remove hood and rear side panels. On models equipped with cab, remove battery cables, remove instrument panel front cover and raise instrument panel as necessary. On models without cab, disconnect battery cables and remove battery. On all models, loosen hose clamps and remove brake reservoir. Remove equalizer pipe from master cylinders, then remove the outlet tubes from master cylinders.

On models not equipped with cab, remove locknuts and eccentric pivot bolts (Fig. 141). On models with cab, loosen clevis locknuts and remove clevis pins (Fig. 142). On all models, unbolt and remove master cylinders.

To disassemble master cylinders, first remove clevis and locknut from push rod on models with cab. Then, on all models, remove boot (13—Fig. 145). Remove snap ring (15) and push rod (14 or 16). Remove gland seal (12), piston (11), spring (10) with retainer, wave washer (9), valve stem (8) and seal (7). Unscrew connector (1) with seal ring (2), then remove equalizer flow valve (3), seal (4) and ball (5).

Clean and inspect all parts and renew complete master cylinder if piston (11) or body (6) is excessively worn or badly scored. A service seal kit is available

for the master cylinder. The kit consists of seal ring (2), seal (4), ball (5), seal (7), wave washer (9), gland seal (12) and snap ring (15). Reassemble by reversing disassembly procedure.

Reinstall master cylinders by reversing removal procedure. Adjust pedal height as outlined in paragraph 121 and bleed air from brake system as in paragraph 122.

POWER TAKE OFF

The power take off used on all models is an independent type. The pto is available as a single speed (540 rpm) unit on Models 385 and 485 or as a dual speed (540 and 1000 rpm) unit on Models 585, 685 and 885. Oil for the pto unit on all models

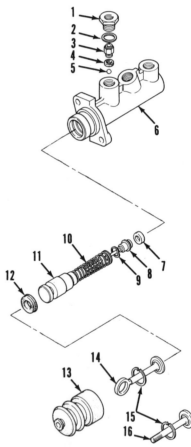

Fig. 145—Exploded view of a brake master cylinder. Other master cylinder is identical. Push rod (14) is used on models without cab and push rod (16) is used on models with cab.

1. Connector
2. Seal ring
3. Equalizer flow valve
4. Seal
5. Ball
6. Body
7. Seal
8. Valve stem
9. Wave washer
10. Spring
11. Piston
12. Gland seal
13. Boot
14. Push rod
15. Snap ring
16. Push rod

Fig. 144—Use special tool IH7847A and slide hammer to remove brake piston.

is furnished from the hydraulic pump through the multiple control valve. From the MCV, the oil is supplied through a cored passage in the rear frame to the 1000 rpm output shaft (Models 585, 685 and 885) or lower pto drive shaft (Models 385 and 485) to actuate the pto clutch. Operation of the pto clutch unit is controlled by a spool type valve located in the MCV housing.

PTO LINKAGE ADJUSTMENT

NOTE: Although some checks and adjustments can be accomplished without removing left rear wheel, most mechanics prefer to remove the wheel to gain working room.

All Models

125. To adjust the pto linkage on models without cab, refer to Fig. 146 and move control lever (1) forward and rearward to check total control spool movement. Total spool movement should be 1.14-1.18 inches (29-30 mm). If not, remove pin (3), loosen locknut (4) and adjust clevis on link (5) to obtain correct total spool movement. Install pin and tighten locknut. Make certain control lever has a small amount of clearance between lever and end of slot in console. If necessary, loosen clamp bolt (2) and while holding linkage in position move lever to adjust. Then, tighten clamp bolt.

To adjust the pto linkage on models equipped with cab, refer to Fig. 147 and move control lever (1) forward and rearward to check total control spool movement. Total spool movement should be 1.14-1.18 inches (29-30 mm). If not, remove pin (7), loosen locknut (6) and adjust clevis on link (5) to obtain correct total spool movement. Install pin and tighten locknut. Check to see that control lever has a small amount of clearance at each end of slot in console. If necessary, remove pin (2), loosen locknut (3) and adjust clevis on link (4) until clearance is obtained. Install pin and tighten locknut.

OPERATING PRESSURE TEST

All Models

126. To check operating pressure, operate tractor until hydraulic fluid is warmed to at least 100° F (38° C). Refer to Fig. 148 and connect a 0-300 psi (0-2000 kPa) pressure gage (1) to the tee fitting (2). Start tractor and operate at low idle speed. Move pto control lever forward to engaged position. Pressure reading

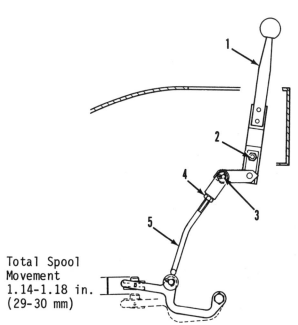

Fig. 146—Pto control linkage used on models not equipped with cab.
1. Control lever
2. Clamp bolt
3. Pin
4. Locknut
5. Adjusting link

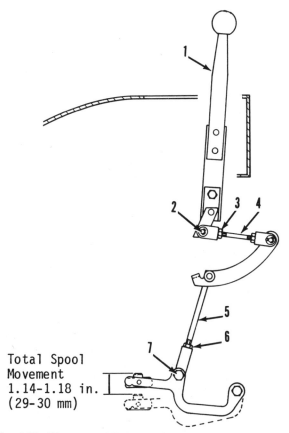

Fig. 147—Pto control linkage used on models equipped with cab.

1. Control lever	
2. Clevis pin	5. Link (lower)
3. Locknut	6. Locknut
4. Link (upper)	7. Pin

on gage should be 260-270 psi (1790-1860 kPa). If reading is correct, there is no problem in hydraulic system of the pto. If reading is not correct, remove test gage and remove pressure line from MCV to tee fitting. Cap tee fitting and connect test gage to the pto outlet union on the MCV as shown in Fig. 149. Operate engine at low idle speed and move pto control lever forward to engaged position.

If pressure was low in first test but is now correct, check for following conditions and correct as required:

  a. Damage to lock ring seals on bottom pto shaft.

  b. Damage to pto shaft bushing in transmission housing.

  c. Damage to seals or piston in pto clutch.

If pressure is low in second test, there is a fault in the hydraulic supply, pto spool, regulator valve or MCV gasket. Refer to paragraphs 25 through 28 for information on hydraulic supply.

With the 0-300 psi (0-2000 kPa) pressure gage still connected to the pto outlet union on the MCV as shown in Fig. 149, move pto control lever rearward into disengaged position. Reading on gage must be zero. A pressure reading on the gage would indicate that spool is not fully up or there is wear on the spool lands.

Move the control lever forward until the pto spool moves downward 0.39-0.43 inch (10-11 mm) (initial engagement position). Pressure reading should be 41-46 psi (283-317 kPa). Record the gage reading, then move control lever fully rearward to disengaged position. Move control lever fully forward to engaged position. Gage reading should be 260-270 psi (1790-1860 kPa). If initial engagement pressure is low but the full engagement pressure is correct, remove the pto spool assembly. See Fig. 150. Inspect the spool lands and if in good condition, add a shim (7). Each shim will increase the initial engagement pressure 3 psi (21 kPa).

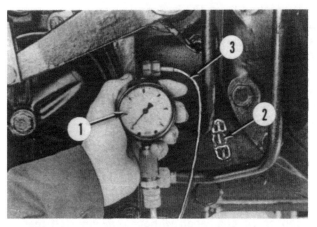

Fig. 149—Pressure gage connected to pto outlet union in multiple control valve to check pressure output.

  1. Pressure gage
  2. Tee fitting
  3. Test gage hose

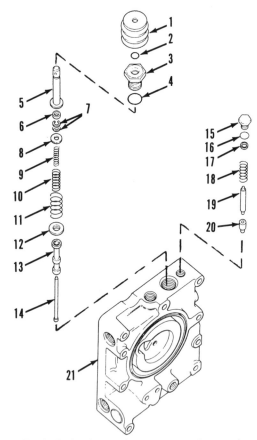

Fig. 150—Exploded view of pto control spool and regulator valve.

1. Boot	
2. "O" ring	
3. Stem guide	13. Spool
4. "O" ring	14. Pin
5. Stem	15. Plug
6. Collar	16. "O" ring
7. Shims	17. Shim
8. Washer	18. Spring
9. Spring (heavy)	19. Pin
10. Spring (light)	20. Regulator valve
11. Return spring	poppet
12. Retainer	21. Multiple control
	valve

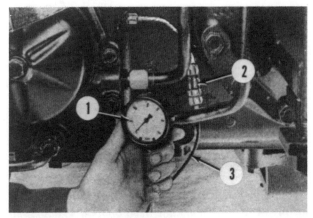

Fig. 148—Pressure gage connected to tee to check pto clutch pressure.

  1. Pressure gage
  2. Tee fitting
  3. Test gage hose

If the initial engagement pressure is correct but the full engagement pressure is low, remove regulator valve (15 through 20). Inspect the regulator valve poppet (20) and seat and if in good condition, add a shim (17). Each shim will increase the full engagement pressure 20-30 psi (138-207 kPa).

NOTE: Repeat pressure tests after adding shims or renewing any parts.

PTO CLUTCH, SHAFTS AND GEARS

All Models

127. R&R AND OVERHAUL. To remove the pto clutch, first drain range transmission and remove clutch cover or transfer case, as equipped. Rotate upper pto output shaft until roll pin is visible as shown in Fig. 151. Drive roll pin from clutch and shaft. Unbolt and remove lower pto shaft cover or seal housing, as equipped, and remove gasket from rear of tractor. Have helper support pto clutch and idler gears, if so equipped. Refer to Fig. 152 and withdraw

Fig. 151—Bottom view of installed pto clutch. Arrow shows location of roll pin which secures clutch to rear pto shaft.

Fig. 152—View showing 1000 rpm output shaft (dual speed) being removed. Lower shaft on single speed is similar.

1000 rpm output shaft (dual speed) or lower drive shaft (single speed). Remove idler gears, if so equipped, and pto clutch as shaft is withdrawn. See Fig. 153.

To disassemble the pto clutch, first use a pair of "C" clamps or a suitable puller and compress the return springs (39—Fig. 154), then remove snap ring (41). Remove clamps or puller and remove clutch back plate (40) and return springs (39). Remove friction discs (37), drive plates (38), piston back plate (36) and brake ring (35). Remove snap ring (44), thrust washer (42) and clutch hub (43). Bump front of clutch cup (31) on bench top and remove piston (32) with "O" rings (33 and 34).

NOTE: Standard duty pto clutch has five friction discs and eight drive plates (two drive plates between each friction disc). Heavy duty pto clutch has six friction discs and five drive plates (one between each friction disc).

Clean and inspect all parts and renew any showing excessive wear or other damage. Pay particular attention to the clutch friction discs and drive plates which should be free of scoring or warpage. Use all new "O" rings, seals and gaskets during reassembly.

Bushing (26—Fig. 154) for the 1000 rpm output shaft (52) or lower drive shaft (22) on 540 rpm unit is in the front of differential compartment. This bushing should be renewed if new shaft (52) or (22) is required or if seal rings (25) have grooved the bushing. When installing bushing (26) bottom it in bore. Position needle bearing (27) to within 1/32-3/32 inch (0.79-2.38 mm) of front face of bushing (26).

Unlock and remove seal rings (25). Remove snap rings (24) and (20) or (50), then press bearing (21) or (51) off rear of shaft (22) or (52) and press gear (23) off front of shaft. Check to see that oil holes are clear and that roll pin is tightly in position in shaft. Use new seal rings and snap rings during assembly.

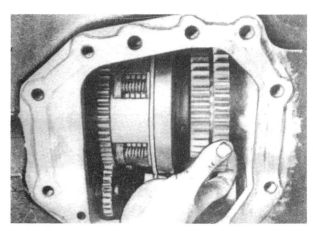

Fig. 153—Remove idler gears and clutch through bottom opening. Idler gears are used only on tractors equipped with front drive axle.

Inspect sleeve (28), roller bearing (29) and idler gear (17), if so equipped, and renew as necessary. Sleeve has a slot which must be aligned with pin in shaft (22) or (52) during assembly.

Reassemble and reinstall pto by reversing disassembly and removal procedures.

To remove the 540 rpm output shaft (8), gear (9), idler shaft (15) and idler gear (17), first remove hy-draulic lift assembly as outlined in paragraph 137 or 138. Remove snap ring (10) and unbolt bearing and seal retainer (2). Withdraw output shaft assembly and remove gear (9) through top opening. Install a cap screw in rear of idler shaft (15), pull idler shaft from rear of frame and remove idler gear (17) from above. Use new "O" ring (14), oil seal and gasket and reinstall by reversing the removal procedure.

1. Shaft cover
2. Bearing & seal retainer
3. Gasket
4. Oil seal
5. Snap ring
6. Washer
7. Ball bearing
8. 540 rpm shaft
9. Gear
10. Snap ring
11. Snap ring
12. Needle bearing
13. Snap ring
14. "O" ring
15. Idler shaft
16. Needle bearing
17. Idler gear
18. Cover
19. Gasket
20. Snap ring
21. Ball bearing
22. Lower drive shaft (540 rpm)
23. Gear
24. Snap ring
25. Seal rings
26. Bushing
27. Needle bearing
28. Sleeve
29. Roller bearing
30. Idler gear
31. Clutch cup
32. Piston
33. "O" ring (inner)
34. "O" ring (outer)
35. Brake ring
36. Piston back plate
37. Friction disc
38. Drive plates
39. Return springs
40. Clutch back plate
41. Snap ring
42. Thrust washer
43. Clutch hub
44. Snap ring
45. Needle bearing
46. Shaft safety cover
47. Bearing & seal retainer
48. Gasket
49. Oil seal
50. Snap ring
51. Ball bearing
52. 1000 rpm shaft

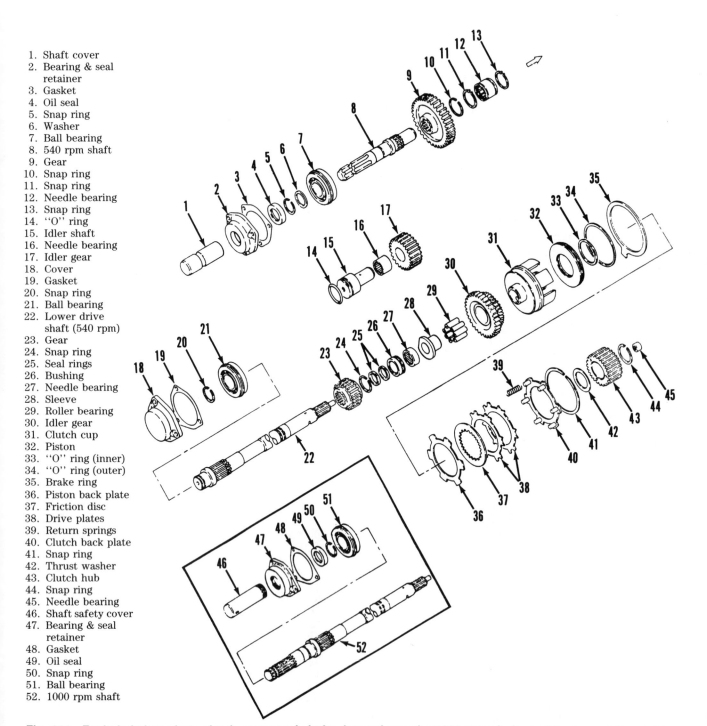

Fig. 154—Exploded view of pto clutch, gears and shafts. Inset shows the 1000 rpm shaft used in place of lower drive shaft (22).

HYDRAULIC LIFT SYSTEM

The hydraulic lift system provides load (draft) and position control in conjunction with the 3-point hitch.

The load (draft) control is taken from the third (upper) link and transferred through a bellcrank on the torsion bar and linkage to this draft control valve located in the hydraulic lift housing. The externally mounted torsion bar is located on the rear side of the lift unit.

The hydraulic lift housing, which also serves as the cover for the differential portion of the tractor rear main frame, contains the work cylinder, rockshaft, valves and the necessary linkage. Auxiliary valves (either one, two or three), on tractors so equipped, are mounted on right side of the lift housing. The pump which supplies the hydraulic system is attached to the multiple control valve (MCV) which is mounted on left side of rear frame. Pump is driven by a gear on the rear of pto driven shaft. The oil used in the hydraulic system is drawn from the rear main frame and through a full-flow filter which is a part of the MCV on left side of rear frame.

TROUBLE SHOOTING

All Models

128. The following are symptoms which may occur during operation of the hydraulic lift system. By using this information in conjunction with the TEST and ADJUST information, no trouble should be encountered in servicing the hydraulic system.

1. Hitch will not lift. Could be caused by:
   a. Faulty main relief valve.
   b. Faulty cushion relief valve.
   c. Internal linkage disconnected.

2. Hitch lifts when auxiliary valve is actuated. Could be caused by:
   a. Unloading valve orifice plugged.
   b. Unloading valve piston sticking.
   c. Unloading valve body assembly not seating or body assembly loose.

3. Hitch lifts very slowly. Could be caused by:
   a. Unloading valve leaking.
   b. Excessive load.
   c. Faulty main relief valve.
   d. Faulty cushion relief valve.
   e. Work cylinder or piston scored or "O" ring faulty.
   f. Flow control valve stuck in slow position.
   g. Faulty pump.

4. Hitch will not lower. Could be caused by:
   a. Main control valve spool sticking or spring faulty.
   b. Drop poppet "O" ring is damaged or drop poppet is sticking.

5. Hitch lowers very slowly. Could be caused by:
   a. Action control valve spool or piston sticking.
   b. Action control valve linkage maladjusted.
   c. Drop poppet valve "O" ring damaged.

6. Hitch lowers too fast in action control zone. Could be caused by:
   a. Improper adjustment of action control valve.

7. Hitch will not maintain position. Could be caused by:
   a. Work cylinder or piston scored or piston "O" ring damaged.
   b. Cylinder cushion relief valve leaking.
   c. Drop poppet check valve ball not seated.
   d. Drop poppet ball seat binding in the drop poppet.

8. Hydraulic system stays on high pressure. Could be caused by:
   a. Linkage maladjusted, broken or disconnected.
   b. Auxiliary valve not in neutral.
   c. Mechanical interference.

9. System stays on high pressure after lifting load, but returns to low pressure after slight movement of position control lever toward "LOWER" position. Could be caused by:
   a. Leak in piston side of unloading valve circuit.

10. Hitch senses with load control in "OFF" position. Could be caused by:
    a. Improper adjustment of load sensing linkage.

TEST AND ADJUST

All Models

129. Before proceeding with any testing or adjusting, be sure the hydraulic pump is operating satisfactorily, hydraulic fluid level is correct and filter is in good condition. All tests should be conducted with hydraulic fluid at operating temperature of at least 100° F (38° C). Cycle system, if necessary, to ensure that system is completely free of air. Remove cab or fenders, fuel tank and seal platform, as equipped, as outlined in paragraph 145 or 144.

130. RELIEF VALVE AND FLOW DIVIDER. To check the pilot relief valve (29—Fig. 155) and flow divider (33), first remove plug (30), then remove spring (32) and spool (33). Reinstall plug (30). Remove pilot relief valve (29) and install a test pilot relief valve from test kit No. 399858R92. Disconnect the

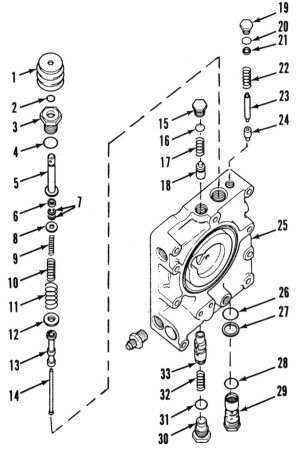

Fig. 155—Exploded view of multiple control valve assembly.

1. Boot
2. "O" ring
3. Guide
4. "O" ring
5. Stem
6. Spacer
7. Shims
8. Washer
9. Spring (heavy)
10. Spring (light)
11. Valve return spring
12. Retainer
13. Valve spool
14. Guide pin
15. Plug
16. "O" ring
17. Spring
18. Oil cooler bypass valve
19. Plug
20. "O" ring
21. Shim
22. Spring
23. Pin
24. Pto pressure regulator valve
25. Multiple control valve housing
26. "O" ring
27. Back-up ring
28. "O" ring
29. Pilot relief valve
30. Plug
31. "O" ring
32. Spring
33. Flow divider spool

power steering supply pipe (directly below hydraulic filter). Cap the pipe and union on multiple control valve with high pressure caps. Disconnect hitch supply line (bottom rear line on multiple control valve) and cap the line. Connect the inlet hose from a flowmeter to the hitch supply union on the multiple control valve. Place outlet hose from flowmeter into transmission oil fill hole and wire in place. Fully open load valve on flowmeter. Start engine and operate at a speed of 2180 rpm for Models 385 and 485, 2300 rpm for Model 585 or 2400 rpm for Models 685 and 885. Record the flow reading. Close load control valve to obtain a reading of 2000 psi (13790 kPa) on pressure gage, then record the flow. Fully open load valve and stop engine. The second flow reading must not be less than 90 percent of the first flow reading. If second reading is too low, check for the following conditions:

1. Low oil level.
2. Clogged filter.
3. Faulty pump.

Remove the test pilot relief valve. Using new "O" rings (26 and 28) and backup ring (27), install original pilot relief valve (29). Start engine and operate at low idle speed. Fully close the load valve on the flowmeter and check the reading on the pressure gage. If pressure reading is not between 2500-2600 psi (17240-17930 kPa), renew pilot relief valve.

Disconnect the flowmeter from hitch supply union on MCV. Remove caps and connect hitch supply line. Remove plug (30) and install original priority flow divider (33) and spring (32), then reinstall plug (30). Remove cap from power steering supply union on MCV and connect inlet hose on flowmeter to the steering supply union. Start engine and operate at low idle speed. Turn load valve in to obtain a reading of 1000 psi (6895 kPa) on pressure gage and record the flow reading. Increase the load to 1600 psi (11030 kPa) and record the flow reading. Fully close load valve and record pressure reading. Fully open the load valve. Increase engine speed to 2180 rpm for Models 385 and 485, 2300 rpm for Model 585 or 2400 rpm for Models 685 and 885. Repeat tests at 1000 psi (6895 kPa), 1600 psi (11030 kPa) and with load valve fully closed. All flow readings should be 2.6-3.1 gpm (9.9-11.8 L/min.). Pressure with load valve fully closed should be 2500-2600 psi (17240-17930 kPa). If flow and pressure readings are not correct, remove and inspect flow divider spring (32) and spool (33). Spring should have a free length of 3.253 inches (82.63 mm) and should test 29.7 lbs. (132 N) when compressed to a length of 2.017 inches (51.23 mm). Check spool (33) and spool bore in housing (25) for excessive wear, scoring or other damage. Spool (33) and housing (25) are a matched set and are renewed only as a new multiple control valve assembly.

131. QUADRANT LEVERS FRICTION ADJUST-MENT. To adjust control lever friction on models equipped with cab, move position control lever fully forward and the draft control lever fully rearward. Adjust locknut on the pivot bolt to just prevent draft lever from moving forward under spring pressure. Move draft control lever fully forward and position control lever fully rearward. Connect a spring scale to knob end of position lever. Adjust locknut on the pivot bolt until a pull of 5-6 lbs. (22-26 N) is required to start lever moving forward. Move position control lever fully forward and the raise rate lever to the rearmost position. Connect a spring scale to knob end of raise rate lever. Adjust locknut on the pivot bolt until a pull of 2-4 lbs. (8.9-17.8 N) is required to start lever moving forward.

To adjust control lever friction on models without cab, move position control lever fully forward and draft control lever fully rearward. Connect a spring scale to knob end of draft control lever. Adjust locknut in pivot bolt until a pull of 2-4 lbs. (8.9-17.8 N) is required to start lever moving forward. Move raise rate lever fully rearward and connect a spring scale to knob end of raise rate lever. Adjust locknut on the pivot bolt until a pull of 2-4 lbs. (8.9-17.8 N) is required to start lever moving forward. Move draft control lever fully forward and position control lever fully rearward. Connect a spring scale to knob end of lever. Adjust locknut on the pivot bolt until a pull of 2-4 lbs. (8.9-17.8 N) is required to start position control lever moving forward.

132. CONTROL LEVER POSITION. On models equipped with cab, set the distance between the centers of the pins on the upper adjusting rod for the draft control lever to 3.40 inches (86 mm). Then, tighten the locknut. Set the distance between centers of pins on the upper adjusting rod for the position control lever to 3.30 inches (84 mm) and tighten the locknut. Move position control lever fully forward and draft control lever fully rearward. Turn the lower adjusting rod until the draft lever is at an angle of 10 degrees forward from vertical, then tighten locknut. Move draft control lever fully forward and position control lever fully rearward. Turn the lower adjusting rod until position lever is at an angle of 15 degrees forward from vertical and tighten locknut. Move raise rate lever fully forward so that lower lever touches the bracket. Remove cotter pin and washer, then turn adjusting rod until the raise rate lever is 11 degrees forward from vertical. Install washer and cotter pin. With position control lever fully forward, move raise rate lever to vertical position. Turn adjusting screw so that the adjusting arm is also vertical, then tighten locknut.

On models without cab, set the distance between the centers of pins on draft adjusting rod to 3.40

inches (86 mm), then tighten locknut. Set the distance between centers of pins on position adjusting rod to 3.30 inches (84 mm) and tighten locknut. Move the raise rate lever to vertical position. Turn adjusting screw until the adjusting screw until the adjusting arm is also vertical, then tighten locknut.

HITCH ADJUSTMENTS

All adjustments must be made with cab or fenders, tank and platform installed.

All Models

133. POSITION CONTROL LEVER. Check the travel of the position control lever. The lever must not touch the console at either end of slot. Make any adjustment to the lever travel at the adjusting rod.

Start engine and operate at 1000 rpm. Raise and lower hitch several times to remove any air in the system. Move draft lever fully forward and raise rate lever rearward into the "FAST" position. Move position control lever rearward to maximum height position. When hitch is at maximum height, measure distance between center of lift link pin and top face of rear frame. See Fig. 156. This distance must be 15.93 inches (404 mm). If measurement is not correct, remove seat and on models with cab, remove the access plate. On all models, unbolt and remove cover plate from hydraulic housing. Refer to Fig. 157 and loosen locknut on draft valve rod. Using a 8 mm wrench on the flats on draft valve rod, adjust the height of the hitch. Looking forward, turn draft valve rod clockwise to decrease or counterclockwise to increase the hitch height. Check to see that outer end of lower lift arms can be lifted by hand at least a further 2 inches (51 mm). If movement is less than 2 inches (51 mm), turn draft valve rod counterclockwise to correct. Tighten locknut.

Move position control lever forward into full lower position, but not into drop speed rate position.

Fig. 156—With hitch at maximum height, distance between center of lift link pin and top face of rear frame should be 15.93 inches (404 mm).

When hitch links have stopped lowering, loosen locknut and turn adjusting screw until it just touches the stop plate in the hydraulic lift housing. See Fig. 158. Tighten locknut.

134. DRAFT CONTROL LEVER. Start engine and operate at 1000 rpm. Move position control lever into full lower position and the raise rate lever rearward into the fast raise position. Move draft control lever fully forward. Use a pry bar to move the draft sensing shaft into the hydraulic housing. Insert the 5/16 inch (8 mm) end of setting spacer (special tool to be made up) between the bracket and end of sensing shaft as shown in Fig. 159. Special tool Fig. 160 can be made up in shop from a 5/16 inch (8 mm) square mild steel bar (6-1/2 inches (165 mm) long. Loosen locknut, then make adjusting rod for draft control lever shorter until the hitch just starts to raise. Then, turn adjust rod 1 to 1-1/2 turns in opposite direction. Tighten locknut. Remove the setting spacer (special tool) and check to see that sensing shaft is contacting the bracket.

135. DROP VALVE SCREW. Use the adjusting wrench (Case special tool No. IH9690) to turn the adjusting screw in the draft valve clevis fully clockwise. See Fig. 161. Turn adjusting screw six revolutions counterclockwise. Place a weight of at least 200 lbs. (90 kg) onto the hitch lower links. Start engine and operate at 1000 rpm. Move raise rate lever rearward into fast raise position and the draft control lever fully forward. Move position control lever fully rearward. If the hitch moves up and down without moving the control levers, stop engine and lower hitch to ground. Turn adjusting screw an additional 1/2 turn counterclockwise, then repeat the test. Start engine and when hitch is at the maximum height and

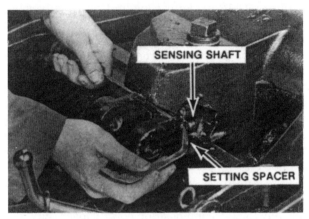

Fig. 159—Insert special setting spacer between third link bracket and sensing shaft.

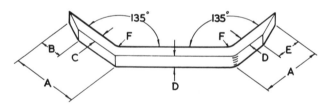

Fig. 160—Special tool to be made up in shop. Refer to text.

A. 1-1/2 inches (38 mm)	D. 5/16 inch (8 mm)
B. 5/8 inch (15 mm)	E. 3/4 inch (19 mm)
C. 1/4 inch (6 mm)	F. 1/2 inch (12 mm) radius

Fig. 157—Turn draft valve rod to adjust hitch height. Refer to text.

Fig. 158—With hitch in lowered position, turn adjusting screw until it just touches stop plate in hydraulic housing.

Fig. 161—Use Case special tool No. IH9690 to adjust drop valve screw. Refer to text.

has stopped moving, stop the engine. Place a mark on a lift arm and the hydraulic housing. Slowly turn adjusting screw clockwise until hitch just starts to lower, then turn the screw 3/4 turn counterclockwise. Remove adjusting wrench and using a new gasket, install cover plate.

UNLOADING AND FLOW CONTROL VALVE

All Models

136. R&R AND OVERHAUL. On models without cab, unbolt and remove seat. On models with cab, remove floor mat, seat and floor plate. Then, on all models, clean dirt from unloading and flow valve and surrounding area. Disconnect inlet and outlet pipes and cap all openings. Unbolt and remove the valve assembly and place on a clean work bench.

To disassemble the unloading and flow control valve, remove plug (1—Fig. 162) with "O" ring (2). Use pliers and remove piston (3) with "O" ring (4). Using special tool IH7846A or equivalent, unscrew unloading valve body (5). Remove unloading valve (6) and spring (7). Remove check valve plug (12) with "O" ring (11), spring (10) and check valve (9). Remove union (17) with "O" ring (16), then withdraw spring guide (15), spring (14) and flow control valve (13).

Clean and inspect all parts for excessive wear or other damage. If flow control valve (13) or housing (8) are scored or damaged, renew complete unloading and flow control valve assembly. Unloading valve (6) and unloading body (5) are available only as a matched set. Check springs (7, 10 and 14) against the following specifications:

Unloading valve spring (7),
  Free length . . . . .1.571 in.
                                      (39.90 mm)
  Test length . . . . .1.260 in.
                                      (32.00 mm)
  Test load . . . . .15 lb.
                                      (67 N)
Check valve spring (10),
  Free length . . . . .0.460 in.
                                      (11.68 mm)
  Test length . . . . .0.400 in.
                                      (10.16 mm)
  Test load . . . . .3.2 oz.
                                      (89 gram)
Flow control spring (14),
  Free length . . . . .1.703 in.
                                      (43.26 mm)
  Test length . . . . .1.200 in.
                                      (30.48 mm)
  Test load . . . . .11 lb.
                                      (49 N)

Lubricate all internal parts with Hy-Tran Plus fluid during reassembly. Install all new "O" rings. Reassemble and reinstall by reversing the disassembly and removal procedures.

HYDRAULIC LIFT HOUSING

Models Without Cab

137. REMOVE AND REINSTALL. To remove the hydraulic lift unit, first remove fenders, fuel tank and platform as outlined in paragraph 144. Remove third link from sensing bracket and disconnect lift links from upper lift arms. Remove remote valve pipes and remove couplers. Cap or plug all disconnected hydraulic lines immediately to prevent entrance of dirt or other foreign material into the system. Disconnect inlet and outlet pipes from raise rate valve and the unloading and flow control valve. Disconnect, unbolt and remove pto shaft and tube assembly. Unbolt and remove plate above transmission. Disconnect and remove any other necessary hydraulic tubes. Unbolt lift housing, attach a sling and hoist, then lift housing assembly from tractor.

Thoroughly clean mounting surfaces, apply Loctite 515 to transmission face and install three new Teflon seals in face of transmission. Then, reinstall hydraulic housing by reversing the removal procedure. Tighten the 3/8 inch cap screws evenly to a torque of 33-37 ft.-lbs. (45.6-51.5 N·m) and the 1/2 inch cap screws

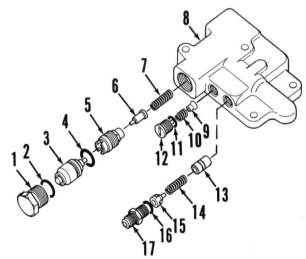

Fig. 162—Exploded view of unloading and flow control valve.

1. Plug	
2. "O" ring	
3. Unloading piston	10. Spring
4. "O" ring	11. "O" ring
5. Unloading body	12. Plug
6. Unloading valve	13. Flow valve
7. Spring	14. Spring
8. Housing	15. Spring guide
9. Check valve	16. "O" ring
	17. Union

to a torque of 80-90 ft.-lbs. (110.6-124.4 N·m). Adjust control levers as outlined in paragraphs 131, 132, 133 and 134.

Models With Cab

138. REMOVE AND REINSTALL. To remove the hydraulic lift housing, first remove the cab as outlined in paragraph 145. Disconnect lower end of adjusting rod for raise rate lever, then disconnect lower adjusting rods from draft and position control levers. Separate adjusting rods from auxiliary valve control levers, if so equipped. Disconnect pto linkage and lift off the control levers box. Disconnect speed transmission and range transmission shift linkage. Disconnect park brake linkage. Disconnect and remove inlet and outlet pipes from raise rate valve. Cap or plug all hydraulic openings immediately. Remove pto tube and cross shaft. Disconnect and remove remote valve pipes and remote couplers. Disconnect and remove pipe from unloading and flow control valve. Disconnect and remove any other necessary hydraulic tubes. Remove third link from sensing bracket and disconnect lift links from upper lift arms. Unbolt lift housing, attach a sling and hoist, then lift housing from tractor.

Thoroughly clean mounting surfaces and apply a 1/8 inch (3 mm) bead of Loctite 515 on the center of the mounting surface of transmission case. Install three new Teflon seals in grooves in face of transmission. Reinstall hydraulic housing over the two locating dowels. Tighten the 3/8 inch cap screws to a torque of 33-37 ft.-lbs. (45.6-51.5 N·m) and the 1/2 inch cap screws to a torque of 80-90 ft.-lbs. (110.6-124.4 N·m). Complete installation by reversing the removal procedure. Adjust control levers as outlined in paragraphs 131, 132, 133 and 134.

CYLINDER AND VALVE UNIT

All Models

139. R&R AND OVERHAUL. To remove the work cylinder and valve assembly, first remove the lift unit from tractor as outlined in paragraph 137 or 138. Turn lift unit upside down on a work bench. Disconnect control linkage from main valve and other linkage. Disconnect sensing spring, then unbolt and remove cylinder and valve assembly from housing as shown in Fig. 163. Disconnect the cylinder lube pipe from the action valve, then unbolt and remove the lube pipe. Unbolt and remove the action valve.

Push inward on spool (4—Fig. 164) to compress spring (3). Remove retaining clip (1), then remove spool and spring from valve body (2). Inspect spool and bore in valve body for scoring or other damage. Spool (4) and body (2) are available only as a matched set. Free length of spring (3) should be 1.705 inches

(43.31 mm) and should test 14.6 lbs. (65 N) when compressed to a length of 0.875 inch (22.23 mm). Lubricate with Hy-Tran Plus oil when reassembling.

To disassemble the control valve, refer to Fig. 165 and remove snap ring (1), plug (2) with "O" ring (3) and spring (4). Remove drop valve poppet with back-up rings (5) and "O" rings (6 and 8), spring (9), ball (10) and drop valve spool (11) with "O" ring (12). Remove snap ring (18) and withdraw draft valve assembly (14 through 23) from valve body (13). Remove actuating rod (21) from draft valve rod (19). Clamp clevis (23) in a soft jawed vise and loosen locknut (20). Unscrew and remove clevis (23). Use a valve spring compressor to compress the spring, then note number of turns required and remove locknut. Carefully release pressure on the spring and remove collar (17) and spring (16) from spool (15). Remove retaining clip (14), then withdraw rod (19) with screw (22) from spool.

Clean all parts in a suitable solvent and dry with compressed air. The draft valve spool (15) and valve body (13) are a matched set and if either part is scored or otherwise damaged, renew both parts. Check actuating rod (21) and renew if not straight. Inspect drop valve poppet (7) and spool (11) for excessive wear or other damage and renew as necessary. Spring (9) should have a free length of 0.920 inch

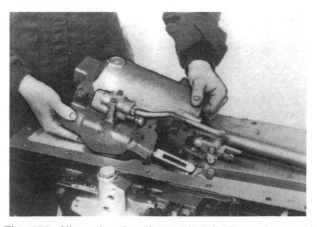

Fig. 163—View showing the work cylinder and control valve assembly being removed from lift housing.

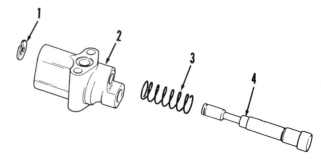

Fig. 164—Exploded view of action valve.

1. Retaining clip
2. Valve body
3. Spring
4. Spool

(23.37 mm) and should test 3.84 lbs. (17 N) when compressed to a length of 0.750 inch (19.05 mm). Free length of spring (4) should be 1.525 inches (38.74 mm) and should test 10 lbs. (45 N) when compressed to a length of 0.640 inch (16.26 mm). Spring (16) should have a free length of 3.174 inch (80.62 mm) and should test 30 lbs. (133 N) when compressed to a length of 1.90 inch (48.26 mm). When reassembling, use all new "O" rings and back-up rings and lubricate them with petroleum jelly. Install locknut as near to original setting as possible.

NOTE: Position of draft valve rod (19) in clevis and adjusting screw (22) in rod (19) must be adjusted after tractor is reassembled.

Lubricate draft valve parts with Hy-Tran Plus oil during assembly. Reassemble by reversing disassembly procedure.

If not already removed, remove cushion (safety) valve (2—Fig. 167) from cylinder (1—Fig. 166). The cushion relief (safety) valve is set at the factory to 2500 psi (17238 kPa). Do not attempt to disassemble or adjust this valve. Bump cylinder (1) on a block of wood and move piston (2) to open end of cylinder. Remove piston (2) with "O" ring (3) and back-up ring (4) from cylinder. Inspect piston and cylinder bore for

scoring or other damage and renew as necessary. Lubricate new "O" ring (3) and back-up ring (4) with petroleum jelly and install on piston so "O" ring is

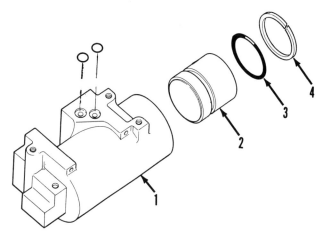

Fig. 166—Exploded view of work cylinder.

1. Cylinder
2. Piston
3. "O" ring
4. Back-up ring

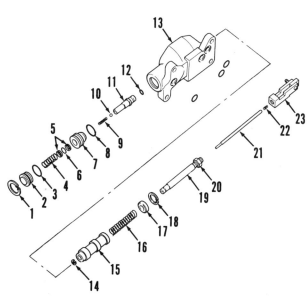

Fig. 165—Exploded view of the lift control valve.

1. Snap ring	
2. Plug	13. Valve body
3. "O" ring	14. Retaining ring
4. Spring	15. Control valve
5. Back-up rings	spool
6. "O" rings	16. Spring
7. Drop valve	17. Collar
poppet	18. Snap ring
8. "O" ring	19. Rod
9. Spring	20. Locknut
10. Ball	21. Actuating rod
11. Drop valve spool	22. Adjusting screw
12. "O" ring	23. Clevis

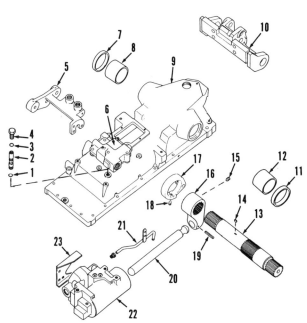

Fig. 167—View of hydraulic lift housing and relative components.

1. "O" ring	
2. Safety valve	13. Rockshaft
3. "O" ring	14. Woodruff key
4. Plug	15. Set screw
5. Raise rate valve	16. Bell crank
6. Unloading & flow	17. Actuating hub
control valve	18. Set screw
7. Oil seal	19. Roll pin
8. Bushing (R.H.)	20. Rod
9. Lift housing	21. Cylinder
10. Sensing bracket	lubrication pipe
11. Oil seal	22. Cylinder
12. Bushing (L.H.)	23. Bracket

ROCKSHAFT

nearest to closed end of piston. Install piston, closed end first, into cylinder.

Use all new "O" rings and reassemble action valve and control valve to cylinder, then reinstall the cylinder and valve assembly into lift housing (9—Fig. 167). Cushion relief (safety) valve (2) can be installed in cylinder through hole in lift housing. Reinstall lift housing as outlined in paragraph 137 or 138.

RAISE RATE VALVE

All Models

140. R&R AND OVERHAUL. To remove the raise rate valve (5—Fig. 167), first remove hydraulic lift housing as outlined in paragraph 137 or 138. Disconnect adjusting rods (5—Fig. 168) from control shafts. Disconnect internal linkage as necessary and remove draft control shaft (25) and position control shaft (6). Unbolt and remove raise rate valve assembly. Remove adjusting arm (21), return spring (20) and plug (19) with "O" ring (18). Remove valve spool (15) with "O" rings (16) and back-up rings (17). Inspect raise rate valve spool and bore in valve body for scoring or excessive wear and renew parts as required.

Use new "O" rings (16) and back-up rings (17) and reassemble and reinstall by reversing the disassembly and removal procedures.

All Models

141. REMOVE AND REINSTALL. To remove the rockshaft (13—Fig. 167), first remove hydraulic lift housing as in paragraph 137 or 138. Then, remove work cylinder and valve assembly as in paragraph 139. Disconnect and remove sensing linkage as required. Disconnect link from actuating hub (17). Remove retaining bolts and washers from both ends of rockshaft. Place marks on lift arms in alignment with timing marks on ends of rockshaft. Remove both lift arms. Loosen set screw (15) in bell crank (16). Loosen set screw (18) in actuating hub (17). Slide actuating hub off the Woodruff key (14) and remove the key. Pull rockshaft (13) out left side of housing and remove actuating hub and bell crank. Remove oil seals (7 and 11) and bushings (8 and 12) as required.

Reinstall by reversing the removal procedure. Install oil seals (7 and 11) with lips facing inward. Lubricate seals with petroleum jelly. Use Loctite 241 on threads of set screws (15 and 18) and tighten securely. Align timing marks and install lift arms. Tighten retaining bolts to a torque of 98-110 ft.-lbs. (133-149 N·m).

1. Position control lever
2. Friction discs
3. Belleville washers
4. Link
5. Adjusting rod
6. Position control shaft
7. Bearing
8. "O" ring
9. Lever
10. Spacer
11. "O" ring
12. Link
13. Spacer block
14. Raise rate valve body
15. Valve spool
16. "O" ring
17. Back-up ring
18. "O" ring
19. Plug
20. Return spring
21. Adjusting arm
22. Raise rate lever
23. Pivot bolt
24. Draft control lever
25. Draft control shaft
26. Pivot bolt

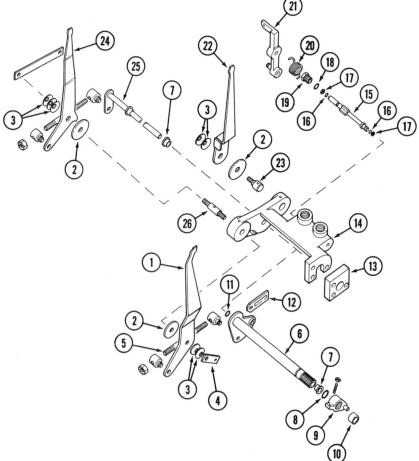

Fig. 168—Exploded view of typical raise rate valve and control lever arrangement.

FILTER, FLUID AND PUMP

All Models

142. Tractors may be equipped with Sunstand or Plessey hydraulic gear type pumps. For information and service procedures, refer to paragraph 26 for the pump and to paragraph 21 for filter change and hydraulic fluid capacity.

AUXILIARY CONTROL VALVE

All Models So Equipped

143. R&R AND OVERHAUL. To remove the auxiliary control valves, apply park brake and block wheels securely. Remove right rear wheel. On models without cab, unbolt and remove slotted panel on right side. On models with cab, disconnect valve linkage. On all models, clean dirt from valves and surrounding area. Disconnect remote pipes from valves, then unbolt and remove end plate and valves as an assembly. Unpin and remove control levers or control links. Remove end plate with three cap screws and separate valves.

> NOTE: Two types of Bosch remote auxiliary valves are used. Bosch No. 0521609005 (Fig. 169) is equipped with two hydraulic check valves. Bosch No. 052169006 (Fig. 170) is equipped with one hydraulic check valve and one mechanical check valve located in the raise work port. To identify the valves, check the number on the plate between the work ports. Disassembly of both valves is similar, except on valves so equipped, the mechanical check valve must be removed first to prevent damaging the valve spool.

To disassemble the valves, refer to Fig. 172 and remove fittings from the raise work port. Bend both ends of retaining clip (33) inward to clear tips on valve retainer (32). Turn retaining clip clockwise to clear valve retainer. Using needlenose pliers, press down on valve retainer and rotate it 90 degrees counterclockwise so the ends fit into large grooves in the

bore. Remove valve retainer (32), large spring (31), washer (30) and small spring (29). Remove retaining clip (33), then remove check valve (27) with ball (28) and actuating pin (26).

On both valves, refer to Fig. 171 or 172 and unbolt and remove bracket (8) with wiper seal (7) and spacer ring (9). Unbolt and remove detent cover (25). With-

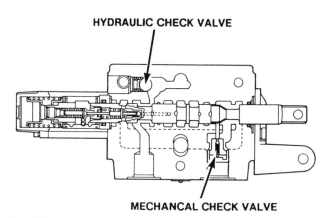

HYDRAULIC CHECK VALVE

MECHANCAL CHECK VALVE

Fig. 170—Sectional view of Bosch No. 0521609006 valve equipped with one hydraulic check valve and one mechanical check valve.

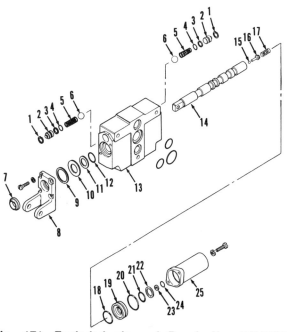

Fig. 171—Exploded view of Bosch No. 052169005 auxiliary valve. Detent assembly not shown.

1. Snap ring
2. Plug
3. Back-up ring
4. "O" ring
5. Spring
6. Check ball
7. Wiper seal
8. Bracket
9. Spacer ring
10. Spacer
11. Back-up ring
12. "O" ring
13. Valve body
14. Valve spool
15. Ball
16. Poppet
17. Spring
18. "O" ring
19. Collar
20. "O" ring
21. "O" ring
22. Back-up ring
23. Shim
24. "O" ring
25. Detent cover

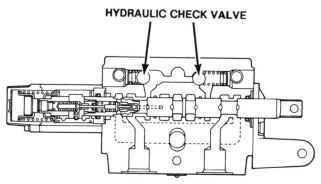

HYDRAULIC CHECK VALVE

Fig. 169—Sectional view of Bosch No. 0521609005 valve equipped with two hydraulic check valves.

draw detent and valve spool assembly from valve body (13). Remove spacer (10), back-up ring (11) and "O" ring from front end of valve body and "O" ring (18) from detent end of valve body. Clamp flats on valve spool in a soft jawed vise. Use an Allen wrench to loosen the detent retaining screw. Remove from vise and hold assembly horizontally close to a bench top. Unscrew detent assembly from valve spool. See Fig. 173. Lay detent assembly aside. Remove spring (17—Fig. 171 or 172), poppet (16) and ball (15) from spool (14). Remove collar (19) with "O" rings (20 and

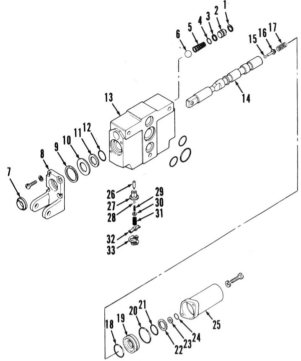

Fig. 172—Exploded view of Bosch No. 052169006 auxiliary valve. Detent assembly not shown. Refer to Fig. 171 for legend except for the following.

26. Actuating pin	30. Washer
27. Check valve	31. Spring
28. Ball	32. Valve retainer
29. Spring	33. Retaining clip

21) and back-up ring (22) from spool. Remove shims (23) and "O" ring from detent assembly.

NOTE: DO NOT attempt to disassemble the detent assembly as no parts are available and special equipment is required for assembly.

To remove the hydraulic check valve(s), press inward on plug (2), remove snap ring (1), then plug with "O" ring (4) and back-up ring (3). Remove spring (5) and check ball (6).

Clean and inspect all parts for excessive wear or other damage. A seal kit containing all seals and "O" rings and repair kits for mechanical check valve (items 26 through 33) and for hydraulic check valve (items 1 through 6) are available for repair. Valve spool (14) and body (13) are a matched set and are not available separately.

Use all new seals and "O" rings and reassemble by reversing the disassembly procedure. When reinstalling auxiliary valves and end plate, tighten the three cap screws to a torque of 12-15 ft.-lbs. (16-20 N·m). Balance of reinstallation is the reverse of removal procedures.

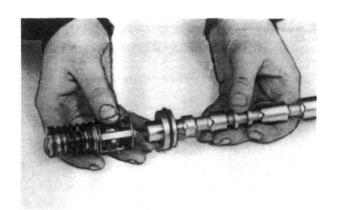

Fig. 173—View of detent assembly removed from valve spool. Do not disassemble detent unit as no parts are available.

FENDERS, FUEL TANK AND PLATFORM

Models Without Cab

144. REMOVE AND REINSTALL. Remove the hood top panel above batteries, then disconnect battery cables. Unbolt and remove the roll over protection system (ROPS). Drain fuel from tank. Disconnect fuel supply and fuel return lines. Disconnect the fuel tank balance line at both sides. Unbolt and remove seat assembly. Disconnect electrical connectors at right and left sides and disconnect the ground leads. Remove the seat deck nuts. Remove knobs and rub-

ber boots from speed transmission and range transmission shift levers. Remove hydraulic control lever and pto lever knobs. Remove screws and remove left and right lower control panels. Remove park brake lever bolts and foot plate to fender bolts. Remove fender mounting bolts at both sides. Attach a sling around tank as shown in Fig. 174 and using a hoist, lift assembly from tractor.

Reinstall the assembly by reversing the removal procedure.

Fig. 174—View showing sling and hoist used to lift off fenders, fuel tank and platform assembly.

CAB

Models So Equipped

145. REMOVE AND REINSTALL. To remove the cab, remove battery cover and disconnect battery cables. Remove cab floor mat and insulation. Unbolt and remove seat. Remove rubber cover from seat platform, then unbolt and remove the access panel. Remove knobs from hydraulic control levers and from speed and range transmission shift levers. Unbolt and remove left console. Disconnect park brake linkage and park brake switch, then unbolt and remove the park brake lever. Remove the four speed and range lever carrier bolts. Unbolt and remove the hydraulic lever operating console. Drive out roll pins and remove auxiliary valve levers. Remove the four hydraulic lever carrier bolts. Disconnect heater hoses. Remove steering column shroud, then unbolt steering support. Unbolt and remove cab floor panel. If so equipped, disconnect power shift electrical connector. Remove hood top rear panel and disconnect wiring harness connectors. Drive out roll pin and disconnect hitch leveling screw extension. Support rear of tractor and remove rear wheels. Remove rear cab center mount bolts. Disconnect air conditioning self sealing connections. Unbolt receiver-drier and tie unit out of the way to transmission. Disconnect differential lock pedal linkage. Disconnect the sev-

en pin electrical connector and disconnect fuel sender wire. Remove plugs from each side of cab and install M16 lifting eyes. Connect a sling and hoist to lifting eyes as shown in Fig. 175. Remove front cab mounting bolts, pads and spacers. Carefully raise cab and remove wiring harness rearward. Remove cab from tractor.

Reinstall cab by reversing the removal procedure. Tighten front and rear cab mounting bolts to a torque of 110 ft.-lbs. (149 N·m). Tighten rear wheel nuts to a torque of 205-220 ft.-lbs. (278-298 N·m).

Fig. 175—Install lift eyes and attach sling and hoist to lift cab from tractor.

NOTES

NOTES

NOTES

NOTES

NOTES

NOTES

NOTES